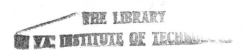

INTRODUCTION TO
BIOELECTRODES

INTRODUCTION TO BIOELECTRODES

Clifford D. Ferris
Bioengineering Program
The University of Wyoming
Laramie, Wyoming

PLENUM PRESS · NEW YORK AND LONDON

Library of Congress Cataloging in Publication Data

Ferris, Clifford D
 Introduction to bioelectrodes.

 Bibliography: p.
 Includes index.
 1. Electrodes. 2. Electrophysiology. I. Title.
QD571.F43   541'.3724   74-19381
ISBN 0-306-30780-4

© 1974 Plenum Press, New York
A Division of Plenum Publishing Corporation
227 West 17th Street, New York, N.Y. 10011

United Kingdom edition published by Plenum Press, London
A Division of Plenum Publishing Company, Ltd.
4a Lower John Street, London W1R 3PD, England

Printed in the United States of America

Preface

This book is the outgrowth of several courses that the author has taught during the last decade at three universities. Most recently, it has served as the principal text for a course offered in the Bioengineering Graduate Program at the University of Wyoming.

The book is designed to fill two needs. For the casual reader who just wants to know something about electrodes, it provides a general overview of the types of electrodes available for different uses. For the student, clinician, and researcher, theories are discussed and practical methods are described.

Both fabrication methods and techniques for use are presented for a variety of electrodes, as well as electrode systems and configurations. The discussion applies to electrodes in both the stimulating and recording modes.

In addition to fabrication and use techniques, there is extensive discussion of various problems associated with electrodes. Attention is directed to electrode polarization (both alternating- and direct-current phenomena), electrical noise, and requirements for backup instrumentation such as electronic amplifiers. A brief treatment of signal analysis and filtering techniques is included to complement the chapter dealing with amplifiers and the discussions of noise.

Each chapter ends with a list of pertinent references so that details not presented in the text may be further explored. In a work of this scope, it is not realistic to list every reference. Those cited have been selected either for very specific information, or because they are review papers with extensive bibliographies. A short, general bibliography appears at the end of the book.

An attempt has been made to effect a balanced treatment of both macro- and microelectrodes, metallic and nonmetallic electrodes.

From time to time, a reference is made to various commercially available electrodes. Such comments only indicate the types of electrodes available

v

and do not constitute an endorsement of specific products. Commercial electrodes may exhibit quite a range of variability and the user is advised to consult the specification and suggested-applications sheets supplied by the manufacturers.

The author wishes to express his thanks to the many graduate students who fabricated and calibrated experimental electrodes for use in preparing this text. Dr. R. Lynn Kirlin, Dr. J. W. Steadman, and Professor R. W. Weeks kindly read selected portions of the text and made helpful suggestions. Sue Watson, Joanne Adachi, and Diane Alexander expertly coped with my difficult calligraphy and other technical matters in the typing and preparation of the final manuscript, and much thanks is owed them. Special thanks are also due Marilyn Larson who kept a busy dean's office on even keel during the preparation of the figures and the revision of the text to final form.

Laramie, Wyoming *Clifford D. Ferris*

Contents

CHAPTER 4

CHAPTER 5

CHAPTER 6

CHAPTER 8

CHAPTER 9

CHAPTER 10

CHAPTER 1

Metallic Electrodes— Introduction

The problem of what type of electrodes to use is one that is critical in biomedical research and experimentation. Electrodes have a dual purpose. They may be used to apply a stimulating signal (excitation) to a physiological system or they may be used to detect the presence of an electric potential in such a system. Electrodes fall into two broad categories: metallic electrodes and fluid-bridge electrodes. The choice of type depends upon the specific application. Initially we direct our attention to metallic electrodes. The present discussion applies to gross electrodes; microelectrodes will be considered separately.

1.1. Electrode Materials

The number of materials from which metallic electrodes can be fabricated is limited because of problems of physiologic toxicity and mechanical strength. The most commonly used materials are the noble metals, platinum and silver in particular, stainless steel, German silver, and tantalum. Platinum is preferred as it possesses good electrical conductivity, appears generally inert to the body, and is strong mechanically if alloyed with a small percentage of iridium. Platinum does not corrode easily and is especially useful for chronically implanted electrodes in such applications as indwelling cardiac pacemakers.

Silver, while an excellent electrical conductor, is soft mechanically and oxidizes easily. It is more sensitive to corrosion than either platinum or gold. Gold is a satisfactory electrode material, although soft, but since it is nearly as expensive as platinum, the latter may as well be used.

Surgical-grade stainless steel is a generally satisfactory electrode material and may be used either internally or externally. It is strong mechanically and does not corrode easily in most applications, cardiac pacing being one exception. It has the advantage of cheapness relative to the cost of the noble metals.

For external applications (skin surface recording), German silver, stainless steel, and aluminum are frequently used. Although it is highly conductive electrically, copper is generally to be avoided as an electrode material in physiological work. Its toxic salts render it hazardous.

Some success with chronic implantation of normally toxic substances has been achieved in fuel cell and hybrid cell biological power source experiments (DeRosa *et al.*, 1968). Generally, the implanting of toxic metals is to be avoided.

The above comments apply in general to electrodes used for recording electric potentials under low current conditions, that is, small to negligible charge flow through the physiological system (electrolyte)–electrode interface. When electrodes are used in the excitation mode for stimulation, other factors may have to be considered. Of major consideration is electrode destruction by electrolytic action. Stainless steel is particularly sensitive to corrosion when used for chronic stimulation. Aside from corrosion of the electrode itself, undesirable metal ions may migrate into surrounding tissues. Rowley (1963) reported a study on stainless steel cardiac pacemaker electrodes in canines. Monophasic stimulation pulses were used, and the positive electrodes were observed to fail in a matter of 24 to 52 days as the result of electrolytic action.

In this two-electrode system, the cardiac tissue behaved electrochemically as the electrolyte in a simple electroplating cell. While stainless steel is passive under recording conditions, it loses its passivity in the chronic, active state (Evans, 1960). Rowley also reported that spectrographic analysis of tissue excised from the positive electrode region showed that metal ions from the electrode had migrated into the myocardial tissue. He showed from a theoretical Faradaic calculation that a 10-mg iron electrode placed in an ionic medium would be electrolyzed in 19 days by a 90-per-min, 1.5-msec-duration, 10-mA unidirectional stimulus. This and other problems associated with electrodes for myocardial and endocardial chronic implantation have been discussed by Greatbach and Chardack (1968). Hallen *et al.* (1965) have also examined the problem.

Electrode destruction of this nature is related to electrode polarization in an ionic environment. Other forms of electrode polarization will be considered in subsequent sections of this book.

Other matters which have to be considered in electrode selection are surface effects. Silver and silver–silver chloride electrodes tend to react with high-protein solutions such as blood. Hubbard and Lucas (1960) have examined ionic charges of glass surfaces and other materials and their possible role in blood coagulation. Pike and Hubbard (1957) reported increased chemical reactivity of certain glass surfaces in buffer solutions. Factors of these sorts are important in selecting glasses and other materials for ion-specific and microelectrodes.

1.2. Electrode Geometry

The size and shape of metallic electrodes varies considerably according to application. For registration of signals from the surface of the body, such as in electrocardiography, disc- or ellipse-shaped electrodes are commonly used. They are usually slightly concave to fit the body contour better. Their size varies from diameters of 2 in. down to subcutaneous needle electrodes. The usual materials are stainless steel and German silver. For registration of electroencephalographic signals, subcutaneous needle probes are frequently employed, although small discs are commonly used. Silver–silver chloride disc electrodes are frequently used as well. There is no particular material or size problem in the recording of body-surface potentials as the signals are summation signals from gross aggregations of cells. Problems do arise in chronic recording, as in the monitoring of astronauts. This subject will be treated in a later chapter. Of major consideration is the preparation of both the electrode surface and the skin surface. Both must be absolutely clean and free from grease (and hair in some cases). Cleaning may be accomplished by using a mixture of acetone and ethyl alcohol. The skin surface should be abraded slightly to remove the dead surface layer. The electrodes and the skin should be moistened with saturated saline (NaCl) solution to insure positive contact. Rubber straps or adhesive tape may be used to apply slight pressure to maintain contact. Electrode paste is generally overrated and is frequently less effective than saline (Ferris *et al.*, 1966).

External stimulating electrodes used for humans and animals are usually concave discs contoured to fit the area of the body to which they are applied. The head and chest are the principal regions for stimulation relating respectively to electroshock therapy and electronarcosis, and cardiac defibrillation. Needles are generally used in neurologic work.

Indwelling electrodes frequently must be shaped to conform to the body geometry. This is especially true when the electrode must fit into a body cavity such as in the recording of electrovaginal potentials. Recording and stimulation at the surface of the heart places many demands upon the electrodes used because of environment and mechanical motion. Usually for cardiac pacing, a rigid electrode is used (Chardack, 1964), although crocheted

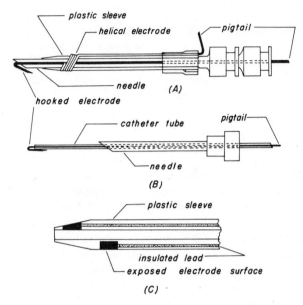

Figure 1.1. Insulated bipolar and hooked electrodes for internal stimulation.

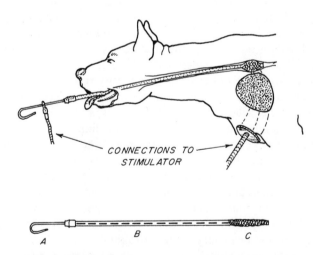

Figure 1.2. Indwelling endoesophageal or endotracheal electrode for cardiac studies. *A*, central stylette; *B*, endotracheal tube; *C*, flexible electrode bulb made from wire mesh.

flexible electrodes have worked quite well (Ferris, 1966). Hook electrodes are used for transvenous pacing (Schwartz *et al.*, 1973).

Some specialized electrodes are shown in Figures 1.1 and 1.2.

1.3. Electrodes for Impedance Measurements

It is frequently necessary to determine the electrical impedance of biological materials and electrolytes. If standard impedance-bridge techniques are used, two electrodes are necessary. The placement of the two electrodes in relation to the biological system determines the sample configuration and related impedance. For surface measurements (such as on the skin surface of an animal), electrodes similar to those used as electrocardiograph pickup electrodes are employed.

When *in vitro* measurements are conducted using special measuring chambers, platinized platinum electrodes are used. This type of electrode, its preparation, and its use are treated in detail in a subsequent chapter.

1.4. Metal Electrodes for Special Purposes

The development of chronically implanted cardiac pacemakers has necessitated an as yet unfinished detailed study of stimulating electrodes. Mechanical and metallurgical considerations are involved. Pacemaker electrodes are frequently implanted directly in the myocardium with two stimulative electrodes required. A major problem relates to the physical design of the electrodes so that they can be sutured to the heart and remain permanently in position. A presently accepted design is the Chardack electrode (Chardack, 1964). Shown schematically in Figure 1.3, it has a plastic housing which protects the top of the electrode and the lead assembly. The electrode is coiled platinum wire. It is sutured in place through holes provided in the plastic housing.

Two metallurgical problems exist. The first is to find a suitable material for the leads which connect to the electrodes, as they must withstand the mechanical flexure of the beating heart. Several stainless steel alloys have been developed. These, available under the trade names "Ethicon" and "Surgaloy," are used in the form of multifilament wires about 0.013 in. in diameter. The second problem relates to electrolysis of the electrodes, as mentioned in Section 1.1. The stimulative signal may be considered as pulsed direct current. As a result, considerable electrolytic action may occur in the electrolyte environment of the heart tissue. It has been reported that stainless steel is eroded very rapidly but that iridium–platinum alloy is quite stable (Rowley, 1963). A possible solution to the problem, in addition to

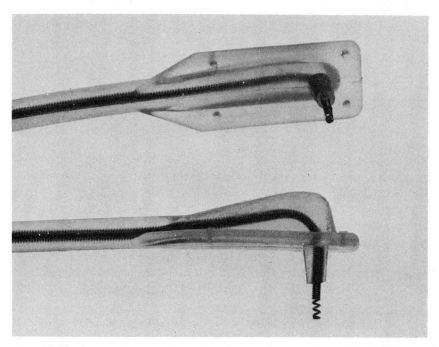

Figure 1.3. Photograph of Chardack electrode used for cardiac stimulation. This type of electrode is now going out of use with the advent of transvenous pacers. (University of Colorado Medical Center photograph.)

using improved electrode materials, is to use a bipolar stimulative pulse rather than a monopolar pulse.

For emergency use when chronic implantation is contraindicated, a single coaxial bipolar electrode is useful. This scheme requires the use of one needle instead of two. The electrode is shown schematically in Figure 1.1. The unit is a No. 18 standard hypodermic needle to which has been added an external insulating sheath and an insulated internal central electrode. It is used in cases of temporary heart block or until corrective procedures can be instituted in the case of permanent block (Ferris and Cowley, 1968).

Chronically implanted stimulating electrodes, whether cardiac or other, are all subject to the problems noted above.

1.5. References

Chardack, W. M., 1964, A myocardial electrode for long-term pacemaking, *Ann. N.Y. Acad. Sci.* **111**(3):893–906.

DeRosa, J. F., Beard, R. B., and Hahn, A. W., 1968, Fabrication and evaluation of cathode and anode materials for implantable hybrid cells, *Proc. 21st ACEMB, Houston*, p. 19.8.

Evans, U. R., 1960, *The Corrosion and Oxidation of Metals: Scientific Principles and Practical Applications*, Edward Arnold, Ltd., London.

Ferris, C. D., 1966, Cardiac resuscitation by electronic stimulators, Final Report, University of Maryland, USPHS Grant HE-4595.

Ferris, C. D., Moore, T. W., and Cowley, R. A., 1966, Frequency and power considerations in the use of alternating current defibrillators, *Bull. School Med. Univ. Maryland* **51**:3.

Ferris, C. D. and Cowley, R. A., 1968, Emergency cardiac pacing system, *Proc. 21st ACEMB, Houston*, p. 22A3.

Greatbach, W. and Chardack, W. M., 1968, Myocardial and endocardiac electrodes for chronic implantation, *Ann. N.Y. Acad. Sci.* **148**(1):234–251.

Hallen, A., Nordlund, S., and Warvsten, B., 1965, Pacemaker treatment in the Adams–Stokes syndroma, *Acta Soc. Med. Upsalien.* **70**:17.

Hubbard, D. and Lucas, G. L., 1960, Ionic charge of glass surfaces and other materials, and their possible role in the coagulation of blood, *J. Appl. Physiol.* **15**(2):265–270.

Pike, R. G. and Hubbard, D., 1957, Increased chemical reactivity of the surface compared with that in the bulk volume of Britton–Robinson universal buffers, *J. Res. Nat. Bur. Std.* **59**(6):411–414.

Rowley, B. A., 1963, Electrolysis—A factor in cardiac pacemaker electrode failure, *Trans. IEEE, PGBME*-**10**(4):176.

Schwartz, M. L., Hornung, J., Helton, W. C., Johnson, F. W., and Nicoloff, D. M., 1973, A new endocardial hook electrode, *Proc. 26th ACEMB, Minneapolis*, p. 2.4.

CHAPTER 2

Alternating-Current Electrode Polarization

2.1. Introduction

Frequently it is necessary to determine the electrical properties of electrolytes and biological materials. Actually, from an electrical point of view, all biological substances can be treated as electrolytes.

The passive electrical properties of a biological material are determined by placing it in a suitable cell which consists of a chamber to contain the material, and if necessary a suspending fluid, and two or more electrodes which contact the tissue–fluid combination. The cell is then connected into a measuring circuit which allows determination of the desired properties. The frequency of the alternating current applied to the cell through the measuring apparatus is varied from 1 Hz to as high as 20 GHz in discrete steps. At each frequency, a determination of electrical impedance is made. In order to cover this wide frequency range, several cells and measuring circuits are required. If the data thus obtained are plotted as a function of frequency, various dips and rises in the resultant curve are noted. Ultimately, these dips and rises can be associated with electrical relaxation times or critical frequencies. The similarity of these quantities to spectral lines in optical spectroscopy has generated the name "alternating-current impedance spectroscopy."

There are many areas in which the passive electrical properties of biological materials are of interest and find application in research work, diagnosis of disease, therapy, and studies of fine structure.

The techniques employed for the determination of the electrical parameters of biological materials depend upon the material and the frequency range of interest.

Biological materials in general behave as electrolytes and may be characterized in terms of resistance (conductance) and capacitance (Schwan, 1957). Since it is usually not necessary to consider inductive properties, measurement is considerably simplified. Before measurement apparatus is discussed, it is first necessary to examine the nature of the measurement and certain extraneous effects.

Since we are interested in the passive properties of biological materials, we may characterize them as general two-terminal impedances. Two possible forms exist for characterizing a two-terminal impedance. One may perform a series determination where the measurement data yield

$$Z_s = R_s + jX_s$$

or the equally valid parallel determination which yields the admittance

$$Y_p = G_p + jB_p$$

If a given sample is subjected to both forms of measurement, then

$$\text{Re}\{Z_s\} = \text{Re}\{1/Y_p\}$$

$$\text{Im}\{Z_s\} = \text{Im}\{1/Y_p\}$$

Figure 2.1 illustrates the two forms in which a sample may be represented.

If a series measurement is made, the data obtained may be converted quite easily to parallel data and vice versa. To do this, the technique shown below may be applied.

Equivalence of the circuits shown in Figure 2.1(a and b) is assumed if

$$\text{Re}\{Z_s\} = \text{Re}\{Z_p\}$$

$$\text{Im}\{Z_s\} = \text{Im}\{Z_p\}$$

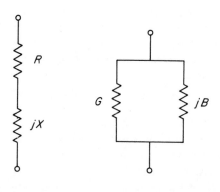

(a)   *(b)*   Figure 2.1. Circuits for equivalence demonstration.

or

$$Re\{Y_s\} = Re\{Y_p\}$$

$$Im\{Y_s\} = Im\{Y_p\}$$

Assume that $Y_p = G_p + jB_p$ has been determined by a measurement and Z_s is to be found; then

$$Z_p = \frac{1}{G_p + jB_p} = \frac{1}{G_p + jB_p} \frac{G_p - jB_p}{G_p - jB_p} = \frac{G_p - jB_p}{G_p^2 + B_p^2}$$

$$= R_p + jX_p = Re\{Z_p\} + j\,Im\{Z_p\}$$

where

$$R_p = \frac{G_p}{G_p^2 + B_p^2} \qquad Re\{Z_s\} = Re\{Z_p\}$$

$$X_p = \frac{-B_p}{G_p^2 + B_p^2} \qquad Im\{Z_s\} = Im\{Z_p\}$$

$$Z_s = R_s + jX_s = Re\{Z_s\} + j\,Im\{Z_s\}$$

Hence

$$R_s = \frac{G_p}{G_p^2 + B_p^2}$$

$$X_s = \frac{-B_p}{G_p^2 + B_p^2}$$

On the other hand, if a series measurement has been made and $Z_s = R_s + jX_s$ is known, then Z_p (or Y_p) may be found in a similar manner:

$$Y_s = \frac{1}{R_s + jX_s} = \frac{1}{R_s + jX_s} \frac{R_s - jX_s}{R_s - jX_s} = \frac{R_s - jX_s}{R_s^2 + X_s^2}$$

$$= G_s + jB_s = Re\{Y_s\} + j\,Im\{Y_s\}$$

$$Re\{Y_p\} = Re\{Y_s\}$$

$$Im\{Y_p\} = Im\{Y_s\}$$

$$Y_p = G_p + jB_p = Re\{Y_p\} + j\,Im\{Y_p\}$$

$$G_p = \frac{R_s}{R_s^2 + X_s^2}$$

$$B_p = \frac{-X_s}{R_s^2 + X_s^2}$$

The transfer equations may be summarized as follows:

$$R_s = \frac{G_p}{G_p^2 + B_p^2}$$

$$X_s = \frac{-B_p}{G_p^2 + B_p^2}$$

to find equivalent series impedance when a parallel determination has been made

$$G_p = \frac{R_s}{R_s^2 + X_s^2}$$

$$B_p = \frac{-X_s}{R_s^2 + X_s^2}$$

to find equivalent parallel admittance when a series determination has been made

Parallel data may be converted to equivalent series data and vice versa by the simple expressions shown immediately above.

Generally it is preferable to carry out an impedance determination in parallel form and later convert the experimental data if series quantities are required. The parallel form of measurement minimizes stray field and stray impedance effects. It also permits a common reference terminal for the resistive and reactive components of the sample measuring circuit.

At low frequencies, the determination of the electrical parameters of electrolytes is complicated by the appearance of an electrode polarization impedance.

2.2. Alternating-Current Electrode Polarization

The discussion which follows applies to both stimulating and recording electrodes. It is presented on an experimental (phenomenological) basis in this chapter. Chapter 3 should be consulted for some of the theoretical considerations involved in defining electrode polarization impedance. Of concern is alternating current at an electrode surface–electrolyte interface. Passage of current at such an interface produces a polarization impedance which is not dependent upon whether the electrode is being used for excitation or recording.

The mechanism of electrode polarization varies slightly for dc and ac currents. (The development which follows makes reference to Figure 2.2.) For the dc case, ions are attracted to the electrodes. An ionic interface layer is established between each electrode and the electrolyte. This process results in an apparent impedance for the solution which is different from the true impedance. The apparent impedance is somewhat higher than the true impedance, and one may consider that the polarization effect at the electrode surfaces contributes a series polarization impedance to the true impedance of the electrolyte. Other effects are discussed in Chapter 3. Electrode boundary

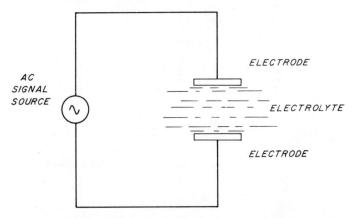

Figure 2.2. Basic model for ac electrode polarization impedance study.

potentials produced by alternating currents are proportional to the current density at the electrode surface and are characterized by an ac Ohm's law behavior. For biological electrolytes, the polarization impedance contains resistive and capacitive reactive components but is primarily capacitive in nature (Schwan, 1951, 1955). Since capacitive reactance, $X_c = 1/(\omega C)$, varies inversely with frequency, one would expect the polarization impedance to decrease uniformly with increasing frequency. Exactly such a frequency dependence is observed experimentally. The actual relation will be developed subsequently. Polarization impedance is strongly dependent upon the type of electrodes used. Since the impedance is a function of current density, an obvious solution is to make the electrode surface area as large as possible to reduce the current density for a given value of current. For practical as well as physical reasons, this is usually undesirable if not impossible. One may increase the effective electrical area without changing the diameter of the electrodes. Platinum has been found experimentally to be the one electrode material which produces the smallest polarization effect. By sandblasting a platinum electrode and then platinizing the electrode (electroplating with colloidal platinum black), its effective surface area can be increased by a factor of as much as 10^4. This area increase is with respect to the polarization elements R_p and C_p. The series value of C_p is increased so that it actually has a much smaller effect upon the measured value of the sample capacitance. In a similar manner, R_p is reduced (Schwan, 1963; Ferris, 1959). Platinization techniques are discussed subsequently.

Electrode polarization is a problem generally at frequencies below 1 kHz and a severe problem below 20 Hz. It is generally not a problem at frequencies above 1 kHz, although it can occur into the megahertz region with certain combinations of samples and measuring cell configurations. The electrical

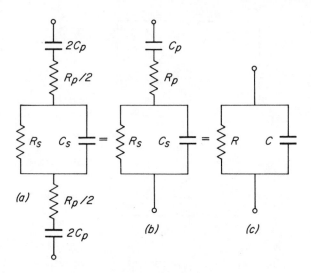

Figure 2.3. Alternating-current polarization impedance: electrical
equivalent of Figure 2.2.

nature of this impedance may be characterized by the electric circuit shown
in Figure 2.3, where

$$C_p = \text{series polarization capacitance}$$
$$R_p = \text{series polarization resistance}$$
$$C_s = \text{true capacitance of electrolyte}$$
$$R_s = \text{true resistance of electrolyte}$$

Figure 2.3a represents the electrical equivalent of the physical system shown
in Figure 2.2. Figure 2.3b is a reduction of the circuit in (a), and (c) represents
the impedance which is measured experimentally (when a parallel impedance
measurement is made) for the cell–electrode configuration. Circuits (b) and
(c) may be shown to be equivalent in the following manner.

Equating impedances, we obtain

$$\frac{1}{(1/R_s) + j\omega C_s} + R_p + \frac{1}{j\omega C_p} = \frac{1}{(1/R) + j\omega C}$$

Rationalizing and equating real and imaginary parts, we get

$$\frac{R_s(1 - j\omega R_s C_s)\omega C_p + R_p[1 + (\omega R_s C_s)^2]\omega C_p - j[1 + (\omega R_s C_s)^2]}{[1 + (\omega R_s C_s)^2]\omega C_p}$$

$$= \frac{R(1 - j\omega RC)}{1 + (\omega RC)^2}$$

$$\frac{R}{1 + (\omega RC)^2} = \frac{\omega R_s C_p + \omega R_p C_p[1 + (\omega R_s C_s)^2]}{\omega C_p[1 + (\omega R_s C_s)^2]} \quad \text{(equating real parts)}$$

$$R = \left(R_p + \frac{R_s}{1 + (\omega R_s C_s)^2}\right)[1 + (\omega RC)^2]$$

$$\frac{\omega^2 R_s^2 C_s C_p + [1 + (\omega R_s C_s)^2]}{[1 + (\omega R_s C_s)^2]\omega C_p} = \frac{\omega R^2 C}{1 + (\omega RC)^2} \quad \text{(equating imaginary parts)}$$

$$C = \left(\frac{1}{R^2}\right)\left(\frac{1}{\omega^2 C_p} + \frac{R_s^2 C_s}{1 + (\omega R_s C_s)^2}\right)[1 + (\omega RC)^2]$$

Or we may use reactance and resistance to find

$$R = \left(R_p + \frac{R_s}{1 + (R_s/X_s)^2}\right)[1 + (R/X)^2]$$

$$X = \left(X_p + \frac{X_s}{1 + (X_s/R_s)^2}\right)[1 + (X/R)^2]$$

Experimental evidence indicates that $(\omega R_s C_s)^{-1} = X_s/R_s$ is always very much greater than unity for $\omega < 100\,\text{kHz}$. In addition, $R_p \ll R_s$, $\omega RC < 1$. Using this information, the equations shown above may be reduced to the following approximations:

$$R \simeq R_p + R_s[1 + (\omega RC)^2]$$

$$X \simeq (X_p + R_s^2/X_s)(X/R)^2$$

$$C \simeq C_s\left(1 + \frac{1}{\omega^2 C_s C_p R^2}\right) = C_s + \frac{1}{\omega^2 C_p R^2}$$

In many experimental situations

$$C_s \ll \frac{1}{\omega^2 C_p R^2}$$

so that

$$C \simeq \frac{1}{\omega^2 C_p R^2}$$

and

$$C_p \simeq \frac{1}{\omega^2 C R^2}$$

and thus

$$R \simeq R_p + R_s(1 + C/C_p)$$

$$C \simeq C_s(1 + C/C_s)$$

Note: If $C \sim C_s$, $\omega RC < 1$ corresponds to $\omega RC_s < 1$. If $C \gg C_s$, $\omega RC \sim \omega RC_p$, and $\omega RC < 1$ corresponds to $1/(\omega C_p) < R$.

Various methods have been proposed for reducing or eliminating the effects of alternating current polarization impedances (Ferris, 1959, 1963; Schwan and Ferris, 1968; Schwan, 1951, 1963). Since the impedance is a function of current density at the electrode surface, an obvious solution is the elimination of current through the probes. Unfortunately, it is impossible to do this in a two-electrode system if one is attempting to measure impedance; however, it can be accomplished with a four-electrode configuration (Ferris, 1959, 1963).

It is obvious from an examination of the expressions presented above for R and C or R_s and C_s that if R_p is large or C_p is small, R_s and C_s cannot be determined with any degree of accuracy from the experimentally measured values of R and C. One solution to this dilemma requires making two measurements. The electrolytic cell is first filled with a specific concentration of saline (usually KCl in H_2O or Ringer's solution), and R and C are measured. The result of this measurement gives the values R_1 and C_1. The saline is emptied from the cell and the cell refilled with an electrolyte of unknown parameters, or the same solution and a semisolid sample may be placed in the cell. The measurement procedure is repeated, yielding the values R_2 and C_2. When the physical situation permits, a simpler technique is to change the electrode spacing by either moving one of the electrodes or by using several fixed electrodes at different positions in the cell (Shedlovsky, 1930). If it is assumed that the polarization impedance remains constant (this may be ascertained by varying the concentration of the saline),* then the true values for the unknown sample may be found from the expressions

$$\left. \begin{aligned} R_1 &= R_{1p} + R_s \\ C_1 &= C_{1p} + C_s \end{aligned} \right\}$$

$$\left. \begin{aligned} R_2 &= R_{2p} + R_u \\ C_2 &= C_{2p} + C_u \end{aligned} \right\}$$

where the bracketed equations represent equivalent parallel bases,

R_{1p} = polarization resistance for saline in cell
       (or for one electrode spacing)

R_{2p} = polarization resistance with sample in cell
       (or for another electrode spacing)

$\left. \begin{aligned} C_{1p} &= \text{polarization capacitance for same condition as } R_{1p} \\ C_{2p} &= \text{polarization capacitance for same condition as } R_{2p} \end{aligned} \right\}$ (parallel basis)

---

*Current through the cell should be maintained constant (constant interface electric current density).

R_s = resistive contribution of saline

R_u = resistive contribution of sample

C_s = capacitive contribution of saline $\left. \right\}$ (parallel basis)

C_u = capacitive contribution of sample

With pure saline, $R_s = R_u$, $C_s = C_u$ when only the electrode spacing is changed. Now for a cylindrical sample

$$R_s = \rho\, d/A$$
$$C_s = \varepsilon_0 \varepsilon_r A/d$$

where

ρ = specific resistance of saline solution, ohm-cm

ε_r = dielectric constant of saline solution

ε_0 = permittivity of free space, F/cm

A = cross-sectional area of cell or electrodes, cm^2

d = electrode separation, cm

The quantities ρ and ε_r are found in standard electrochemical tables for a given saline solution concentration, so that R_s and C_s are known quantities.

If

$$R_{1p} = R_{2p} = R_p$$

and

$$C_{1p} = C_{2p} = C_p$$

then

$$R_1 = R_p + R_s$$
$$S_1 = S_p + S_s$$
$$R_2 = R_p + R_u$$
$$S_2 = S_p + S_u$$

where S_i denotes a reciprocal capacitance or elastance since capacitors in series add as an inverse relation. Thus

$$R_1 - R_2 = \Delta R = R_s - R_u + R_p - R_p$$
$$S_1 - S_2 = \Delta S = S_s - S_u + S_p - S_p$$

and

$$R_u = R_s - \Delta R$$
$$S_u = S_s - \Delta S$$

The expressions shown immediately above indicate how R_u and S_u may be determined using two measurements. The various capacitances and resistances which appear above are series values. This means that if a parallel measurement is conducted, the data must be converted to the equivalent series form and then the expressions shown below can be used:

$$R'_u = (G'_s - \Delta G')^{-1}$$
$$C'_u = C'_s - \Delta C'$$

where R'_u and C'_u are the corrected parallel values.

In many cases, R_p or C_p may have values such that the differences $R_1 - R_2$ or $S_1 - S_2$ are small differences of two very large numbers. Such a situation results in a severe loss of accuracy in ΔR or ΔS. When this condition occurs, other measurement techniques must be used.

2.3. Linear and Nonlinear Behavior

Whenever a metallic electrode is placed in contact with an electrolyte, a dc potential difference can be detected across the interface between the metal and the electrolyte. If an alternating current is passed through the electrode surface, the dc potential difference is amplitude modulated as shown in Figure 2.4. The relation between the degree of modulation and surface density of the alternating current is found to be linear as long as the ac potential does not negate the dc potential at any point in the ac cycle. Within this linear range, the polarization impedance is virtually independent of current density. The modification of the boundary potential is most pronounced for ac and broad-spectrum signals. If we define V_0 as the dc boundary

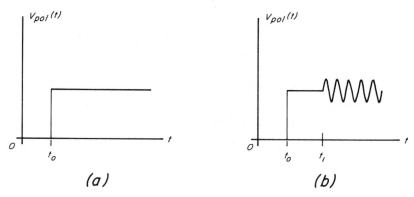

Figure 2.4. Electrode polarization potentials. (b) Illustrates dc electrode polarization with superimposed ac polarization potential. Electrode–electrolyte contact is made at time $t = t_0$ and ac potential is applied at time $t = t_1$.

potential and a phasor current $\dot{I}$ is passed through the interface, then experimentally the following relation may be written for the modifying potential at the interface:

$$\dot{V} = a_1\dot{I} + a_2\dot{I}^2 + \ldots = \dot{Z}_p\dot{I}$$

$$V_{\text{total}} = V_0 + \dot{V}$$

where the a's are determined experimentally and where $\dot{Z}_p$ is the ac polarization impedance. From this relation, we see that the electrode polarization impedance has both linear and nonlinear terms. Experimentally it has been shown to contain the elements of resistance (R) and capacitance (C) (Schwan, 1951) (see Figure 2.3). The linear and nonlinear ranges of $\dot{Z}_p$ are determined by a threshold current $\dot{I}_\tau$ which is a function of a given electrode–electrolyte interface. At current levels below $\dot{I}_\tau$ ($\dot{I} < \dot{I}_\tau$), $\dot{Z}_p$ decreases linearly as the frequency (ω) of the applied current is increased. In the nonlinear range, $\dot{Z}_p$ decreases as $\dot{I}$ increases for ω held constant, as shown in Figure 2.5.

Initial studies of the nonlinear behavior of ac electrode polarization impedance were reported by Schwan and Maczuk (1965). In summary, their

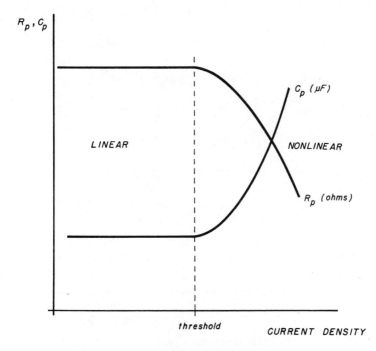

Figure 2.5. Behavior of C_p and R_p as a function of current density at the electrode–electrolyte interface.

work showed that (1) the polarization impedance $\dot{Z}_p$ decreases as a function of current density $\dot{J}$ at the electrodes above a certain threshold $\dot{J}$; (2) $\dot{Y}_p$ is a linear function of $\dot{J}$; and (3) the polarization capacitance is described by

$$C_p = af^{-m} + b\dot{J}f^{-1}$$

Additional measurements (Ferris and Mellman, 1967) were carried out using 0.1M KCl and the following electrodes: stainless steel, bright platinum, sandblasted platinum, platinized platinum; two-parallel-disc electrodes, and disc-and-opposing-point electrodes. The impedance bridge used was limited in the amount of current which it could supply to the electrolytic cell. By using one point electrode, it was possible to increase the current density by several orders of magnitude. The results were consistent with Schwan and Maczuk's prior work.

In addition, the following relationships were found: (1) The factor m is a function of $\dot{J}$ and its value in the linear region differs from its value in

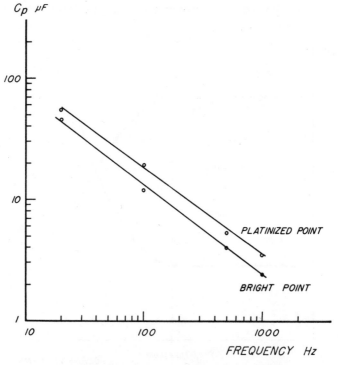

Figure 2.6. Polarization capacitance C_p as a function of frequency for disc electrode and opposing point (platinum electrodes). 10 mA current. (Disc electrode ~ 6.25 cm^2.)

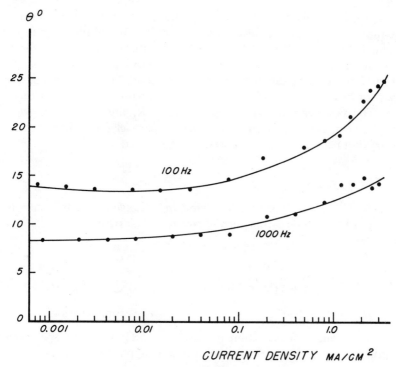

Figure 2.7. Phase angle θ as a function of current density for bright platinum point and disc electrode configuration.

the nonlinear region (this is true for C_p and R_p). (2) The phase angle

$$\theta = \tan^{-1}(\omega R_p C_p)$$

generally varied directly as current density. It remained constant only for platinized platinum. For stainless steel θ increased strongly with frequency. This means that the simple form of Fricke's law (presented in Section 2.4)

$$\theta = m\pi/2$$

is not applicable. This is consistent with the fact that m varies. (3) Since $\dot{Z}_p$ is proportional to current density, sandblasting or platinizing electrodes should reduce $\dot{Z}_p$, as the effective electrode area is increased. It was found, based on limited work, that sandblasting alone increased $\dot{Z}_p$. This requires further study. Platinizing did increase effective area, a result widely published in the literature. Figures 2.6 and 2.7 summarize these results.

Electrode polarization impedance depends upon the electrode material and its surface treatment, the electrode area, the electrolyte, the temperature, and other factors. Platinum electrodes are found to be the best and can be made more effective by the surface deposition of colloidal platinum from a platinum chloride solution (Kohlrausch, 1897). With large electrodes (surface area greater than 1 mm^2), a current density of 1 mA/cm^2 can be tolerated at 1 kHz and linear operation of the electrodes assured. The maximum current density for linear operation is directly proportional to frequency, as noted above. The polarization impedance elements, R_p and C_p, as shown in Figure 2.3, can be characterized in series or parallel form. The choice is arbitrary, but usually the series representation is used.

2.4. Fricke's Law

Fricke (1932) developed the following linear range relations to describe the components of alternating-current electrode polarization:

$$C_p = \frac{\omega^{-m}}{V_0 \Gamma(1 - m)\cos(m\pi/2)} \qquad (\Gamma \text{ function})$$

$$R_p = V_0 \omega^{m-1} \Gamma(1 - m)\sin(m\pi/2) \qquad 0 < m < 1$$

$$\dot{Z}_p = R_p + (j\omega C_p)^{-1} \qquad (\text{series configuration})$$

A more general formulation by Jaron *et al.* (1967, 1969), specifically for cardiac pacemaker electrodes, but of general applicability, gives

$$C_p = C_0\{1 + \omega\gamma^{1-\alpha}\cos(1 - \alpha)\pi/2\}^{-1}$$

$$R_p = (\gamma^{1-\alpha}/C_0)[\sin(1 - \alpha)\pi/2]\omega^{1-\alpha}$$

where C_0, γ, α are experimentally determined parameters.

The potential difference developed across $\dot{Z}_p$ in response to an applied-current step function [$i(t) = 0, t < 0$; $i(t) = I_0, t > 0$] is

$$v(t) = (I_0/C_0)\{t + \gamma^{1-\alpha}t^\alpha[\Gamma(1 + \alpha)]\}$$

A simplified form of Fricke's law states that electrode polarization may be characterized by an electrical phase angle θ which is defined according to the relation

$$\tan\theta = \omega R_p C_p$$

The angle θ is frequency independent provided that C_p changes with frequency according to the relation

$$C_p = C_0 \omega^{-m}$$

If m is frequency independent, then θ may be expressed as follows:

$$\theta = m\pi/2 \qquad (\text{Fricke's law in simplified form})$$

In the experimental situation, the polarization capacitance usually does not vary directly as a power of frequency and Fricke's law holds only approximately, as θ varies from 30° to 45° in the normal experimental situation. Variations in m range from 0.06 to 0.87 (Fricke, 1932) depending upon electrode material and the electrolyte used. For platinized platinum electrodes with physiological saline as the electrolyte,

$$C_p \sim \omega^{-m} \qquad 0.3 \leq m \leq 0.5$$

$$R_p \sim \omega^{-0.5}$$

In many experimental situations, because of inadequacies in impedance bridge resolution, either C_p or R_p cannot be determined. In such situations, Fricke's law can be used to estimate the unknown quantity. Normally $\tan \theta$ varies between 0.5 and 1.0. Under ideal conditions, if one plots on a double logarithmic basis the electrode polarization capacitance C_p (assuming that C_p is known) as a function of natural frequency, the straight-line plot shown in Figure 2.8a will be observed. The exponent m is derived from the plot as indicated. If the plot is not quite a straight line (solid portion in the figure), then Fricke's law may be in error for the associated measurement. Errors can range from small percentages up to 50%.

As an example of how Fricke's law may be applied, let us suppose that we have made a parallel measurement of electrolyte capacitance over the frequency range from 1 Hz to 100 kHz. We have found a measured value of capacitance C_m for a parallel configuration as depicted in Figure 2.3c (shown simply as C rather than C_m in Figure 2.3c). The experimental plot of C_m versus frequency is shown in Figure 2.8b. Let us assume that the electrodes used have an effective surface area of 100 cm^2, and that $R_s = 160 \, \Omega$ approximately. The variation in R is too small to observe R_p. We would like to estimate R_p by using Fricke's law.

If we extrapolate the data to infinite frequency, where $C_p = 0$, then we see that the true sample capacitance $C_s = 0.001 \, \mu F$. If we now plot C_p versus frequency on a double logarithmic basis, we obtain the result shown in Figure 2.8c. The slope of this line is 0.85 and is the coefficient m which appears in the expression of Fricke's law. Normally m varies from 0.3 to 0.5 for platinum electrodes with a saline electrolyte. For the hypothetical sample selected, $m = 0.85$.

An initial rough calculation from the graph of Figure 2.8b indicates that $\omega R C_m < 1$, so that we may use the approximation relations

$$R \simeq R_p + R_s[1 + (\omega R C_m)^2] \sim R_p + R_s$$

$$C_m \simeq C_s + \frac{1}{(\omega R)^2 C_p}$$

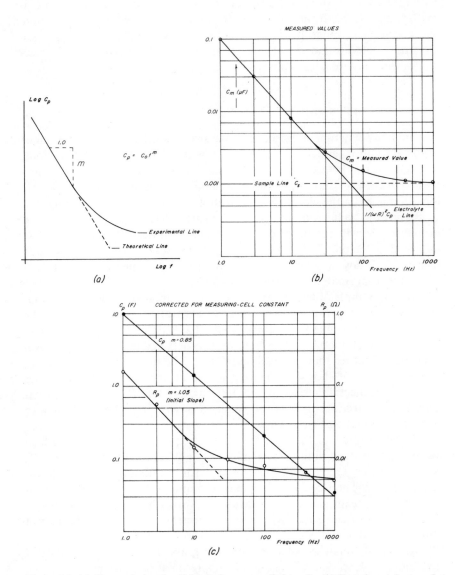

Figure 2.8. (a) Slope relation for Fricke's law as applied to C_p. (b) Experimental curve for electrolyte capacitance determination. (c) Plot of C_p and R_p after Fricke's law has been applied.

TABLE 2.1. Table for Fricke's Law Calculation

Frequency, Hz	$C_m, \mu F$	C_p, F	ωC_p	R_p, Ω
1	0.1	10	6.28	0.662
3	0.03	3.8	18.8	0.221
10	0.008	1.4	88	0.058
30	0.0027	0.55	104	0.040
100	0.0015	0.2	126	0.033
400	0.0011	0.063	158	0.026
1000	0.00103	0.033	207	0.020
10 k	0.001	—	—	—
∞	0.001	0	—	0

Experimentally it is known that $R_p \ll R_s$; thus $R \sim R_s$ and

$$C_p \simeq \frac{1}{(\omega R_s)^2 (C_m - C_s)}$$

To find the polarization resistance as a function of frequency, we use Fricke's law as follows:

$$\theta = m\pi/2$$

$$= 0.85\pi/2 = 0.425\pi = 76.5°$$

$$\tan \theta = \tan 76.5° = 4.16$$

$$\tan \theta = \omega R_p C_p = 4.16$$

$$R_p = 4.16/\omega C_p$$

The numerical values are shown in Table 2.1 and are plotted in Figure 2.8c.

The variation in the slope of R_p versus ω results from assuming $\tan \theta$ as a constant. This indicates the error which can be introduced by the assumption of constancy. In this case, however, we have no other means for estimating R_p.

2.5. Electrode Platinizing Technique

A technique for effectively increasing the electrical area of an electrode without increasing its mechanical dimensions is platinization. Colloidal platinum black is plated onto the metallic electrode surface. The surface granularity thus produced can increase the effective electrode surface by as much as 10,000 times, with an equivalent reduction in current density at the interface. The granularity produces more points at which current lines can leave the electrode surface, or sites where charge can depart from the surface. The technique is best applied to platinum electrodes, but may also be employed with gold, nickel, and stainless steel electrodes. Alloys such as brass tend to poison the platinizing solution. Colloidal gold may be plated onto gold electrodes to give the same effect. Platinizing solution as developed by Kohlrausch (1897) is commercially available (A. H. Thomas Co., Philadelphia, Pa.).

The platinizing technique is somewhat of an art and rather complex if one is to assure repeatable results. The electrodes to be treated must be absolutely free from grease and other contaminants, and should be sandblasted before plating. A bright platinum electrode is used as the anode, and the electrode to be plated as the cathode. Best results are obtained with plating-current densities of the order of $10 \, \text{mA/cm}^2$. Improved coatings are obtained by interchanging the anode and cathode connections for short periods of time. The regimen is somewhat tricky and is discussed below. Platinization effectively increases C_p on the series basis; thus it decreases the term $(j\omega C_p)^{-1}$, and decreases R_p so that $\dot{Z}_p$ is decreased.

When pairs of electrodes, such as would be used in an impedance bridge measuring cell, are platinized, the electrodes, after preparation, should be shorted together in a saline solution and left for several days so that the residual potential difference produced by plating can be equalized.

The preparation of platinized electrodes consists of depositing colloidal platinum (platinum black) upon the working surfaces. This is accomplished by the use of conventional electroplating techniques. The electrode to be platinized is connected as the cathode, and a piece of pure platinum sheet forms the anode of an electrolytic cell. Kohlrausch and Holborn formula platinizing solution is employed as the electrolyte. Its chemical composition is: platinum chloride (H_2PtCl_2) 3% dissolved in 0.025% lead acetate solution. This platinizing solution may be obtained in 2-oz bottles from the Hartman–Leddon Co. in Philadelphia or from the Arthur H. Thomas Co., also of Philadelphia.

Prior to platinization, the electrodes must be cleaned and the surface prepared. The electrode surface should first be sanded with a fine emery paper and water to remove previous colloidal platinum coatings. Boiling in *aqua regia* may be necessary. The electrode surface is then sandblasted with American Optical Co. M180 emery, followed by washing with distilled water. The water wash is followed by three rinses: (1) in acetone, $(CH_3)_2CO$; (2) in a solution of two parts by volume of 95% ethyl alcohol, C_2H_5OH, one part acetone, and one part methyl alcohol, CH_3OH; (3) in absolute ethyl alcohol. This concludes the preplatinizing surface treatment. The electrodes are now placed in a platinizing cell (Figures 2.9 and 2.10) which is rinsed with spent platinizing solution and filled with new platinizing solution. One should note that the platinizing cell contains two sections. This permits the simultaneous platinizing of a matched pair of flat electrodes. A current of 10 mA/cm^2 is passed through each section of the cell for 25 min. The cell is then emptied, the electrodes rotated 180°, the cell refilled, and the plating carried out for an additional 25 min. Rotation of the electrodes prevents the formation of a thickness gradient of the coating. A good coating should have a fine-grained, velvety black appearance.

Figure 2.11 illustrates the effect of platinizing time upon the series electrode polarization capacitance with current density as a parameter. Very low current densities, while providing a good surface, require a prohibitively long plating duration. On the other hand, large current densities cause flaking off of the deposited platinum particles. A current density of 10 mA/cm^2 appears to be the optimum value. A 0.9 % saline (NaCl) solution is taken as the electrolyte for the C_p determination with an electrode separation of 1.0 cm.

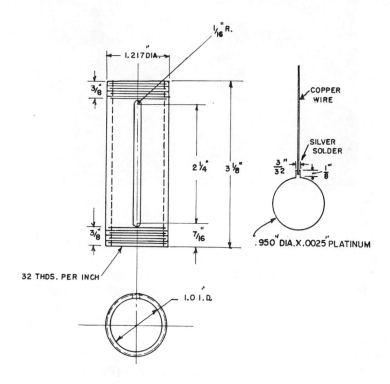

MATERIAL: POLYSTYRENE EXCEPT AS NOTED

Figure 2.9. Construction diagram for a platinizing cell.

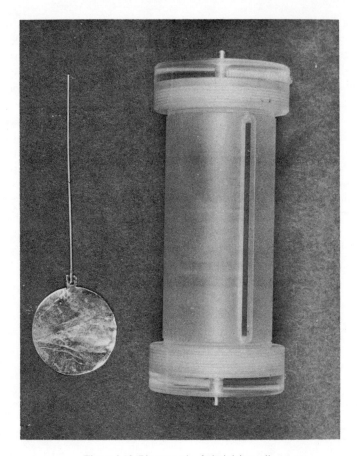

Figure 2.10. Photograph of platinizing cell.

After the electrodes have been prepared, they should be placed in a suitable holder which is filled with distilled water or physiological saline solution. The electrodes are then connected to form a short circuit. The electrodes should remain this way for several days until the charge produced by the platinizing procedure has decayed. Figure 2.12 illustrates the decay of the dc polarization potential (as a result of the platinizing procedure) with time.

A few general comments concerning the care of platinized electrodes are in order. The electrodes should be stored in distilled water. Before use, they should be allowed to stand for 30 minutes in contact with the material to be investigated. If the electrodes have not been damaged during use, re-platinizing is unnecessary. The electrodes should be removed from the

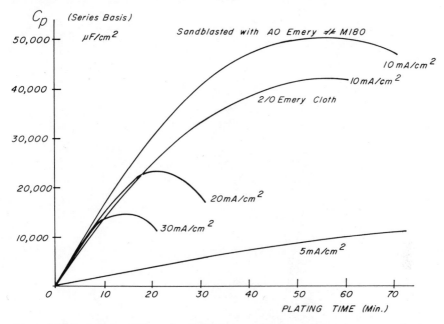

Figure 2.11. Dependence of electrode polarization capacitance on plating time, current density, and surface treatment. Electrode area = 5 cm² (J. Maczuk in Ferris, 1959).

measuring cell, washed with distilled water and ethyl alcohol, and replaced in distilled water. If grease accumulates on the electrodes, it should be removed immediately by washing with ether. Electrodes should be re-platinized if grease cannot be removed or if electrode surfaces are damaged or generally worn down. If a heavy accumulation of grease forms, the electrodes should be placed in boiling *aqua regia* for five minutes.

2.6. Zero-Current Techniques

Platinization of electrodes is not a panacea. It extends the lower frequency range over which electrolyte measurements can be made, and reduces the effects of ac electrode polarization impedance. The only way in which the polarization impedance can be eliminated is to make measurements in a manner that eliminates the current density at the electrode–electrolyte interface.

Two techniques are available for making impedance measurements on a zero-current basis. These are illustrated in Figures 2.13 and 2.14. The simplest method is the basic four-electrode system shown (Ferris, 1963). The specimen to be measured forms one arm of an electronic half-bridge

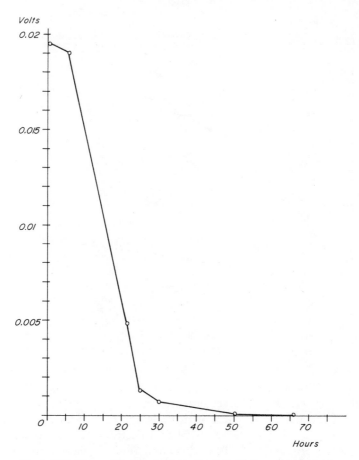

Figure 2.12. Decay of dc polarization potential with time after platinizing. Two platinum electrodes, 6.25 cm² each, platinized for 50 minutes at 10 mA/cm². Cell electrolyte during decay period was NaCl solution; ac cell resistance, 41.4 Ω at 1 kHz.

and supports a series current. The measuring circuit consists of two virtually zero-current potential probes. Bridge balance is achieved when the decades R_v and C_v are adjusted so that the voltage across this combination is equal in magnitude and phase to the voltage developed across the potential electrodes of the measuring cell. The voltage divider is required to produce an effective voltage gain of unity in the first amplifier.

With the system shown, a resistance resolution of 1 part in 100,000 is possible at 1 kHz, which degenerates to 1 part in 1000 at 1 Hz. System noise and detector inadequacies account for loss in resolution. Capacitance

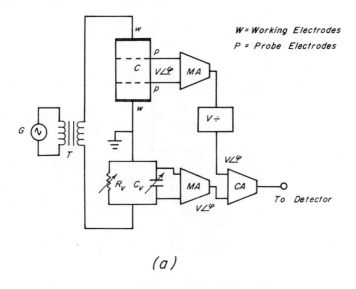

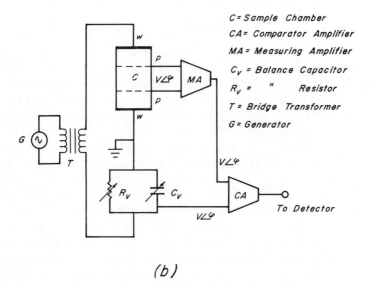

Figure 2.13. (a) Balanced amplifier four-electrode half-bridge system. (b) Simplified system where *MA* is set for unity gain. This system is susceptible to stray capacitances unless care is taken in its use.

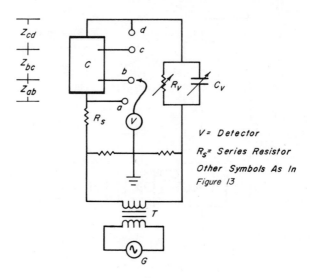

Figure 2.14. Passive bridge zero-current measuring system.

resolution is governed by the relation

$$(\Delta R/R)/(\Delta C/C) = \omega RC$$

where R and C are the true values of the sample.

This system has one shortcoming. Unless the two amplifiers are very carefully matched so that no relative phase shift exists, capacitance measurements are meaningless, although resistance determinations can be made. If the relative phase shift is small, one may substitute a parallel RC circuit for the potential electrodes of the cell and rebalance for null after the initial balance of R_v and C_v. The settings of the new RC decades are the true values for the electrolyte. The system has the advantage that direct readings are obtained with minimum time and effort. Other limitations are described in the literature (Hill and Schwan, 1968). The design of the sample cell is somewhat critical, and details have been published (Schwan and Ferris, 1968). The amplifiers used must have high common-mode signal rejection.

A circuit which avoids the need for a differential amplifier with high common-mode rejection is shown in Figure 2.14 (Schwan and Ferris, 1968). The null detector is a single-ended grounded amplifier which acts as a bridge null detector. Different parts of the total sample impedance are balanced against other parts of the sample in series with the variable RC network. Let the subscripts b and c represent the impedance values of the RC network which apply for the different balance equations when the detector is connected across points b and c:

$$\text{(at } b) \qquad R_s + Z_{ab} = Z_{bc} + Z_{cd} + Z_b$$

$$\text{(at } c) \qquad R_s + Z_{ab} + Z_{bc} = Z_{cd} + Z_c$$

The difference of b and c yields

$$Z_{bc} = \tfrac{1}{2}(Z_c - Z_b)$$

where Z_{bc} characterizes the sample impedance between electrodes b and c without the effect of the electrode polarization impedance. This system, however, is not direct reading, is slow to balance, and has other limitations (Schwan and Ferris, 1968).

In addition to passive impedance measurements, four-electrode techniques find application in active plethysmography as well.

Figure 2.15 illustrates an improved four-electrode measuring system which utilizes commercially available operational amplifiers (Ferris and Rose, 1972). The sample chamber used with this system is shown schematically in Figure 2.16.

It should be pointed out that high-input-impedance measuring systems such as illustrated in Figure 2.13 have certain limitations. In connecting a sample chamber to the electronics, very short lengths of nonshielded wire should be used to minimize stray shunt capacitances. Coaxially shielded cables should not be used because of their inherent high capacitance, usually of the order of 20 pF/ft. Generally, measurements should be carried out in a screen room or Faraday-shielded area to reduce pickup of electrical noise and 60-Hz hum. Even very small shunting capacitances (of the order of 1 or 2 pF) across the input terminals of the operational amplifiers (Figures 2.13 and 2.15) can significantly reduce the input impedance of the system. The "zero-current" property of the system is then destroyed. Since shunt capacitance is involved, the effect is thus frequency-dependent and is less of a problem at low frequencies than at high frequencies.

2.7. A Final Note: Circuit Models, pH and pO₂

In this chapter, the circuit models which have been proposed to represent ac polarization impedance are quite simple. They do, however, give a good fit to experimentally determined data. Esthetically, the simple models are appealing and relatively easy to relate to the physical processes of charge migration, diffusion, and convection described in Chapter 3. More sophisticated models could be proposed, but they would not prove any more useful experimentally than the simple circuits.

Transmission-line representations as well as other models derived from network synthesis techniques have been proposed to represent $\dot{Z}_p$ (Damaskin,

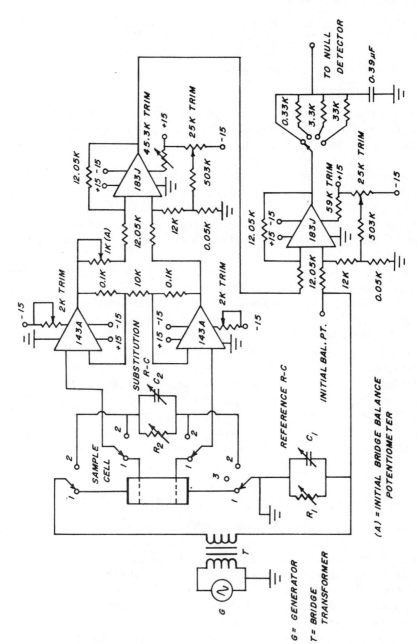

Figure 2.15. Circuit schematic for four-electrode half-bridge system. See Ferris and Rose (1972) for details of operation.

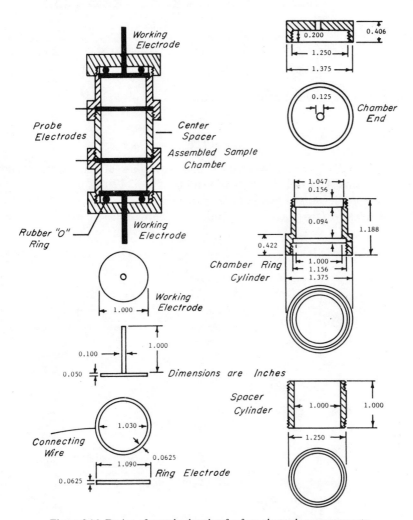

Figure 2.16. Design of sample chamber for four-electrode measurements.

1967; Filanovskaya, 1966; Pollack, 1973). Generally speaking, these are cumbersome in practical use. Some general discussion, however, is now presented to indicate other methods for characterizing ac polarization impedance phenomena. Rate-sensitive processes are examined briefly as they relate to $\dot{Z}_p$.

The rate law for a simple oxidation–reduction process may be expressed as

$$\text{oxidant} + ne \underset{k_2}{\overset{k_1}{\rightleftarrows}} \text{reductant}$$

where

$$ne = \text{number of electrons transferred}$$

$$k_1, k_2 = \text{rate constants}$$

The rates of the forward and reverse reactions are given by

$$\vec{v} = k_1 C_o$$

$$\grave{v} = k_2 C_r$$

where C_o and C_r are the respective concentrations of the oxidized and reduced forms. A model system is shown in Figure 2.17.

A theory developed by Frumkin *et al.* (1952) indicates that the cathodic current density J is given by the difference between the current densities

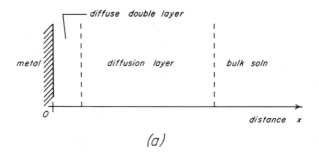

(a)

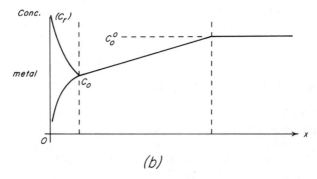

(b)

Figure 2.17. (a) Model for interface between a metal electrode and an electrolyte solution. (b) Variation of the concentration of a reactant in the diffuse double layer and the diffusion layer. C_o and C_o^0 are the surface and volume concentrations of the substance (oxidant). C_r is the concentration of the reductant.

associated with the forward and reverse reactions according to the relation

$$J = \vec{J} - \bar{J} = nF(k_1 C_o - k_2 C_r)$$

$$= nF\left[k_1^\circ C_o \exp\left(-\frac{\alpha nFE}{RT}\right) - k_2^\circ C_r \exp\left(\frac{(1 - \alpha)nFE}{RT}\right)\right]$$

where E is the electrode potential with respect to a hydrogen reference electrode (see Chapter 5), and the superscript $^\circ$ on the rate constants refers to voltage-independent values. F is Faraday's constant, T is Kelvin temperature, R is the universal gas constant, and α is the transfer coefficient as developed by Delahay (1965). Thus the rate constants depend upon electrode potential according to the relations

$$k_1 = k_1^\circ \exp\left(-\frac{\alpha nFE}{RT}\right)$$

$$k_2 = k_2^\circ \exp\left(\frac{(1 - \alpha)nFE}{RT}\right)$$

where k_1° and k_2° are rate constants not dependent upon electrode potential. At some value of electrode (cathode) potential E_e°, the rate constants k_1 and k_2 will be equal such that

$$k_1 = k_2 = k^\circ \quad \text{for } E = E_e^\circ$$

Thus a single rate constant k° may be employed where

$$k^\circ = k_1^\circ \exp\left(-\frac{\alpha nFE_e^\circ}{RT}\right) = k_2^\circ \exp\left(\frac{(1 - \alpha)nFE_e^\circ}{RT}\right)$$

This leads to a new definition for current density and the two original rate constants:

$$J = nFk^\circ\left\{ C_o \exp\left(-\frac{\alpha nF\Delta E}{RT}\right) - C_r \exp\left(\frac{(1 - \alpha)nF\Delta E}{RT}\right)\right\}$$

$$k_1 = k^\circ \exp\left(-\frac{\alpha nF\Delta E}{RT}\right)$$

$$k_2 = k^\circ \exp\left(\frac{(1 - \alpha)nF\Delta E}{RT}\right)$$

$$\Delta E = E - E_e^\circ$$

Starting from the principles outlined above, Damaskin (1967) has developed a theoretical and experimental treatment of ac electrode polarization impedance. His circuit model (p. 73), developed on a series basis, is

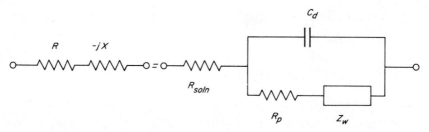

Figure 2.18. Series circuit models for an electrochemical cell. R and X are the simple series resistance and reactance, respectively, in the elementary model. R_{soln} is the bulk solution resistance, C_d is the double layer capacitance, R_p is the polarization resistance, and Z_w is the Warburg impedance.

shown in Figure 2.18. The model represents the electrolyte as a series impedance rather than the parallel impedance developed earlier in this chapter. The impedance components of this model are described by

$$R = R_{soln} + \frac{R_p + \beta\omega^{-\frac{1}{2}}}{(C_d\beta\omega^{\frac{1}{2}} + 1)^2 + \omega^2 C_d^2(R_p + \beta\omega^{-\frac{1}{2}})^2}$$

$$-jX = -j\left\{\frac{\omega C_d(R_p + \beta\omega^{-\frac{1}{2}})^2 + \beta\omega^{-\frac{1}{2}}(\omega^{\frac{1}{2}} C_d\beta + 1)}{(C_d\beta\omega^{\frac{1}{2}} + 1)^2 + \omega^2 C_d^2(R_p + \beta\omega^{-\frac{1}{2}})^2}\right\}$$

$$Z_w = \beta\omega^{-\frac{1}{2}} - j\beta\omega^{-\frac{1}{2}} \quad (\beta \text{ is defined below})$$

Z_w is the Warburg impedance and carries the name of its investigator (1899, 1901). C_d is the double-layer capacitance (Figure 2.17). Experimentally, it is measured by applying a voltage step, $Vu(t)$, through a series resistance, to an electrolytic cell. The system is modeled as a total resistance R_t in series with a double-layer capacitance C_d. The response charging current, i_c, is monitored and C_d is derived from the experimentally observed relation

$$i_c = (V/R_t)\,e^{-t/R_t C_d}$$

$$R_p = \frac{RT}{n^2 F^2 k^\circ C^\circ}$$

is the polarization resistance, and

$$\beta = \frac{\sqrt{2}RT}{n^2 F^2 C^\circ \sqrt{D}}$$

where D is the diffusion constant associated with the diffusion layer (Figure 2.17 and Chapter 3), and C° is the equilibrium concentration of the substance being reduced.

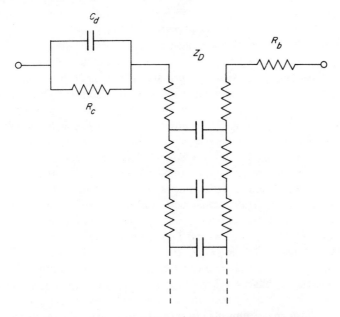

Figure 2.19. Pollack transmission-line model for interface impedance of a metal electrode in contact with an electrolyte. R_c is the charge transfer resistance, C_d is the double layer capacitance, R_b is the bulk resistance of the electrolyte, Z_D is the driving point impedance of uniform, infinite, homogeneous RC transmission line.

Pollack (1973), on the other hand, has chosen a transmission-line representation for the interface impedance, as shown in Figure 2.19. C_d is the double-layer capacitance as described above. The parallel combination of R_c and C_d is used to represent the rate process for the transfer of charge between the metal electrode and the immediately adjacent electrolyte layers (see Figure 2.17). For small potential drops across the interface,

$$R_c \sim 0.026/J_0 A$$

where J_0 is the "exchange current density," A is the active cross-sectional area of the electrode, and 0.026 is the 300°K value for the "temperature voltage" (analogous to a p–n semiconductor junction).

The driving-point impedance Z_D of an infinite homogeneous RC transmission line is used to model the diffusion processes. The magnitude of Z_D is pH sensitive in an experimental situation. R_b represents the resistance of the bulk electrolyte.

Recent studies by DeRosa and Beard (1973) have indicated that ac electrode polarization impedance is sensitive to pH and pO_2. In their experimental work, they used polished platinum electrodes, platinized platinum electrodes, and porous platinum black electrodes. Solid and porous palladium electrodes were also studied.

The porous platinum black electrodes were fabricated by compressing platinum black powder under high pressure (DeRosa *et al.*, 1971).

Experiments were conducted using parallel plate electrodes immersed in 0.001N KCl at 25°C and exposed to atmospheric oxygen. The porous electrodes, by experimental measurement of C_p and R_p, were found to have higher effective areas than even platinized platinum electrodes, that is, higher series-measured C_p and lower series-measured R_p.

Additional studies indicated that ac electrode polarization in the nonlinear range appears to depend upon several rate-determining reactions at electrode surfaces. Noble metal electrodes exhibit polarographic phenomena in the presence of oxygen, when current passes the electrode–electrolyte interface (see Chapters 3 and 4, and Damjanovic, 1969). Hence the interface impedance is sensitive to the pO_2 of the electrolyte.

Variation of the pH of the electrolyte showed, for the nonlinear range, that $\dot{Z}_p$ on the series basis was lowest for very high and very low values of pH. Series C_p and R_p exhibited their lowest and highest values, respectively, for $pH = 7$. Thus the maximum pH effect (maximum series $\dot{Z}_p$) occurs in neutral solutions, which is the basis for most biological measurements.

Hence two rate-determined processes have been identified as contributors to ac polarization impedance associated with aqueous solutions: pH and pO_2.

More will be said about electrode polarization in Chapter 3 and in Section 4.4.

2.8. References

Damjanovic, A., 1969, Mechanistic analysis of oxygen electrode reactions, in *Modern Aspects of Electrochemistry, Vol. 5* (J. O'M. Bockris and B. E. Conway, eds.), Plenum Press, New York.

Damaskin, B. B., 1967, *The Principles of Current Methods for the Study of Electrochemical Reactions*, McGraw-Hill Book Co., New York.

Delahay, P., 1965, *Double Layer and Electrode Kinetics*, Interscience, New York.

DeRosa, J. F., Beard, R. B., Koener, R. M., and Dubin, S., 1971, Porous cathode for implantable power generating electrodes, *Proc. 24th ACEMB, Las Vegas*, p. 276.

DeRosa, J. F. and Beard, R. B., 1973, Electrode polarization studies on solid and porous platinum and palladium, *Proc. 26th ACEMB, Minneapolis*, p. 11.

Ferris, C. D., 1959, Theory, design, and application of a four-electrode system for impedance measurements of biological materials at very low frequencies, *University of Pennsylvania, Interim Report, HEW Grant USPH H 1253(C6)*.

Ferris, C. D., 1963, Four-electrode electronic bridge for electrolyte impedance determinations, *Rev. Sci. Instr.* **34**(1):109–111.

Ferris, C. D. and Mellman, S., 1967, Nonlinear behavior of a-c electrode polarization impedance, *Proc. 20th ACEMB, Boston*, p. 15.3.

Ferris, C. D. and Rose, D. R., 1972, An operational amplifier 4-electrode impedance bridge
for electrolyte measurements, *Med. Biol. Eng.* **10**:647–654.

Filanovskaya, T. P., 1966, On the origin of the anomalous dispersion of dielectric parameters of
living tissue in the range of low radio frequencies, *Biofizika* **11**(2):278–281.

Fricke, H., 1932, On the theory of electrolytic polarization, *Phil. Mag.* **14**:310.

Frumkin, A. N., Bagotskii, V. S., Iofa, Z. A., and Kabanov, B. N., 1952, *Kinetics of Electrode
Processes*, Moscow Univ. Press, Moscow.

Hill, P. L. and Schwan, H. P., 1968, Relative performance of different four-electrode bridges,
Proc. 21st ACEMB, Houston, p. 19.7.

Jaron, D., Schwan, H. P., and Geselowitz, D. B., 1967, A mathematical model for the polarization impedance of cardiac pacemaker electrodes, *Proc. 20th ACEMB, Boston*, p. 15.2.

Jaron, D., Briller, S. A., Schwan, H. P., and Geselowitz, D. B., 1969, Nonlinearity of cardiac
pacemaker electrodes, *Trans. IEEE, BME*-**16**(2): 132–138.

Kohlrausch, F., 1897, Ueber platinierte Electroden und Widerstands bestimmung. *Ann.
Physik Chem.* **60**:315–328.

Pollack, V., 1973, An equivalent diagram for the interface impedance of metal electrodes,
Proc. 26th ACEMB, Minneapolis, p. 14.

Schwan, H. P., 1951, Electrodenpolarization und ihr Einfluss auf die Bestimmung dielektrischer
Eigenschaften von Flüssigkeiten und biologischem Material, *Z. Naturforsch.* **6b**(3):121–
129.

Schwan, H. P., 1955, Electrical properties of body tissues and impedance plethysmography,
Trans. IRE, PGME-**3**:32–46.

Schwan, H. P., 1957, Electrical properties of tissue and cell suspensions, in *Advances in Medical
and Biological Physics*, Vol. 5 (J. H. Lawrence and C. A. Tobias, eds.), Academic Press,
New York.

Schwan, H. P., 1963, Determination of biological impedances, in *Physical Techniques in
Biological Research*, Vol. 6 (W. L. Nastuk, ed.), Academic Press, New York.

Schwan, H. P., 1966, Alternating current electrode polarization, *Biophysik* **3**:181–201.

Schwan, H. P. and Ferris, C. D., 1968, Four-electrode null techniques for impedance measurement with high resolution, *Rev. Sci. Instr.* **39**(4):481–485.

Schwan, H. P. and Maczuk, J. G., 1965, Electrode polarization impedance: limits of linearity,
Proc. 18th ACEMB, Philadelphia, p. 24.

Shedlovsky, T., 1930, *J. Am. Chem. Soc.* **52**:1806.

Warburg, E., *Ann. Physik* **67**:493(1899); **69**:125(1901).

CHAPTER 3

Electrode Polarization and Related Phenomena

In Chapter 2, we approached alternating-current electrode polarization impedance from the phenomenological point of view, which parallels the historical development of this subject. Before we embark upon descriptions of electrochemical cells, ion-specific electrodes, and potentiometric techniques, it is necessary to discuss some of the electrochemical processes that occur at the interface between a solid electrode surface and a contacting electrolyte.

By choice, the treatment here will be brief and incomplete. The processes which take place are complex and depend upon many factors. Two recent books have been entirely devoted to investigations of this subject (Adams, 1969; Newman, 1973). In the present chapter, we simply identify several phenomena and indicate how they relate to electrode use. A detailed discussion is well beyond the scope of this text, and the reader who requires more extensive information is referred to the texts cited above.

3.1. Some Basic Definitions

In Chapter 1, we described, in a rather broad sense, various classes of electrodes. In the study of voltammetry a more specific definition is required. Voltammetry is the study or measurement of the current–voltage relations which exist at the face of an electrode immersed in a solution which contains electroactive species. For our purposes, the solution will be an aqueous electrolyte.

We are primarily concerned with solid electrodes, and throughout this text this will mean, for the most part, platinum, gold, silver, and stainless steel. A *working electrode* now may be defined as any electrode through which net charge flows; hence, it is sustaining a net current through some sort of electrochemical cell. In voltammetric studies, working electrodes may be stationary, rotating, or vibrating in space. The bathing electrolyte may be static, stirred, or flowing. Thus we can have a wide variety of mechanical conditions for an electrode–electrolyte system. In most biological work, there is little relative motion at the interface between an electrode and electrolyte. In blood-flow measurement by electromagnetic flowmeter techniques, stationary electrodes in a flowing electrolyte are involved.

3.1.1. Polarization

For the remainder of this chapter, we will be concerned with electrode surfaces through which a net charge is flowing. Hence no distinction will be made between stimulating or recording electrodes. When charge flow occurs, the electrode has an electric potential which is different from its equilibrium value (no charge flow). The nonequilibrium potential is called an *over-potential* and the electrode is said to be *polarized*. In the strict electrochemical sense, there are no nonpolarizable electrodes (Adams, 1969), although silver–silver chloride electrodes are popularly termed nonpolarizable.

Normally, an electrode system consists of an active or working electrode, an indifferent or reference electrode, and an electrolyte with which the electrodes are in mutual contact. Define V_r as the potential of an electrode which is sustaining a net anodic or cathodic reaction. Let V_e be the electrode's equilibrium value (no net reaction). Thus the total overpotential of the electrode (total polarization potential), V_T, is defined by

$$V_T = V_r - V_e$$

V_T has three additive components:

$$V_T = V_o + V_c + V_a$$

where

$$V_o = \text{ohmic overpotential}$$

$$V_c = \text{concentration overpotential}$$

$$V_a = \text{activation overpotential}$$

The ohmic overpotential is defined by a linear voltage–current relationship at the electrode (Ohm's law). It is derived from a cell in which the electrolyte resistance R is finite with current I passing through the cell. Hence a simple IR drop occurs.

The concentration overpotential results from continued electrolysis, which causes the ionic concentrations at the electrode surface to differ from those in the bulk solution. See Kolthoff and Lingane (1952) for a detailed analysis. V_c is the principal overpotential in voltammetric measurements.

The activation overpotential is related to rate-determined charge-transfer processes. V_a is related to the irreversibility of electrodes, while V_o and V_c result from natural charge flow.

Contributing factors to these potentials are the mass-transfer processes of migration, diffusion, and convection. These matters will be taken up in Section 3.3.

3.2. The Basic Voltammetric Measurement

Figure 3.1 illustrates the basic experimental system for determining the voltage–current curves for an electrode–electrolyte system. The working electrode is frequently gold or platinum and the reference electrode is a conventional saturated calomel half-cell. The current I in the cell is varied by adjusting the sliding contact of the variable resistor. The cell potential is then recorded to obtain voltage–current (voltammetric) curves as shown in

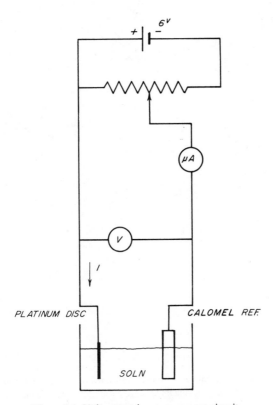

Figure 3.1. Voltammetric measurement circuit.

Figures 3.2 and 3.3. It is essential that a high-input-impedance voltmeter, such as an electronic instrument with 11 MΩ input impedance, be used. If a low-input-impedance instrument is used, part of the recorded current will be that drawn by the voltmeter and not the cell current. Depending upon the cell configuration, either a milliammeter or microammeter is used to measure current. These meters have high internal resistances, and this produces a voltage drop across the meter when current exists. This is the reason it is necessary to monitor the voltage directly across the cell.

The curves shown in Figures 3.2 and 3.3 were obtained using a bright platinum disc electrode (3.37 cm^2 surface area/side) and a calomel reference electrode (Beckman 39412) in an electrolyte containing Fe^{+++} and Fe^{++} and a background electrolyte of 1M H_2SO_4. The solution was approximately 0.05M in Fe^{+++} and Fe^{++} for the curve shown in Figure 3.2. Ferric and ferrous sulfate were used to make the solution. A very dilute concentration (~ 0.0001M) of Fe^{+++} and Fe^{++} was used to obtain the curve in Figure 3.3.

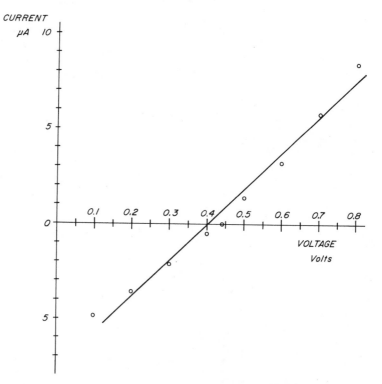

Figure 3.2. Voltage–current curve for high-concentration electrolyte. (Engineering convention rather than electrochemical convention has been used for current and voltage senses.)

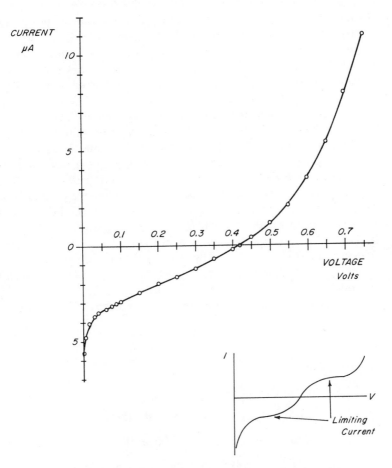

Figure 3.3. Voltage–current curve for low-concentration electrolyte.

The background electrolyte was unchanged. For the first curve, the electrolyte was not stirred; for the second curve, it was stirred.

The basic bulk-solution Nernst equation for this system is

$$E = E° + 0.059 \log[Fe^{+++}/Fe^{++}]$$

where

E = potential of platinum electrode referred to the standard calomel reference

$E°$ = potential of the Fe^{+++}–Fe^{++} system in 1M H_2SO_4 under standard conditions

Now, if the bulk concentrations in the solution are changed by oxidative or reductive titration, E will vary. At zero current, $E = 0.44$ V from the experimental measurement. Figure 3.2 presents a nearly linear relation, which indicates that for high bulk concentrations the electrolyte can be replaced by a fixed resistor. Had the solution been stirred, a more nearly straight line would have resulted. Experimentally, the curve slope is ~ 53 kΩ. A 1-kHz impedance bridge measurement of the system yielded ~ 58 kΩ.

The curve shown in Figure 3.3 results from electron transfer at the electrode surface. Thus the Fe^{+++}–Fe^{++} concentration ratio is altered at the surface, and we are not concerned about the bulk situation. Initially, electrons are transferred to the Fe^{+++} species, and the platinum electrode acts as a working cathode (when $E < 0.44$ V). When $E > 0.44$ V, the platinum electrode acts as an anode and electrons are removed from the Fe^{++} species. The curve is frequently called a polarogram or *polarization curve*.

In the ideal case, shown as the insert in Figure 3.3, the limiting current regions occur when diffusion is the only mass transport mechanism operative. The extreme portions of the curve result from the reduction of hydrogen and the evolution of oxygen (Adams, 1969, Chapter 1).

The basic voltammetric reactions may be summarized in this way:

overall reaction:

$$\text{oxidant} + ne \rightleftarrows \text{reductant}$$

component reactions:

$$(\text{oxidant})_{bulk} \longrightarrow (\text{oxidant})_{electrode}$$

$$(\text{oxidant})_{electrode} + ne \longrightarrow (\text{reductant})_{electrode}$$

$$(\text{reductant})_{electrode} \longrightarrow (\text{reductant})_{bulk}$$

Other component reactions may be present but are ignored.*

3.3. Mass Transfer Considerations

The basic mass transfer processes of interest are migration, diffusion, and convection. We will examine some simple relations in this section. Detailed treatment is beyond the scope of this book. Initially we consider stationary planar electrodes immersed in an unstirred electrolyte. A general treatment is given by Delahay (1954), and a practical approach by Adams (1969).

3.3.1. Migration

Charge migration results from forces exerted by an electric field on charged particles in solution. If we have a parallel-planar electrode con-

*Electron charge is implicit.

figuration with a potential difference V applied between the electrodes, then in an isotropic medium, neglecting fringing effects, the electric field $\mathbf{E}$ between the plates is given by

$$\mathbf{E} = (V/S)\mathbf{a}_x$$

where S is the separation between the electrodes, and it is assumed that the field is acting in the x direction ($\mathbf{a}_x$ is the unit vector). The Coulomb force $\mathbf{F}$ on a charged particle of charge Q is given by

$$\mathbf{F} = Q\mathbf{E}$$

Equating mechanical (Newton's second law) and electrical (Coulomb's law) forces and neglecting charge interactions and body forces such as gravity,

$$\mathbf{a}_x M \, d^2x/dt^2 = Q\mathbf{E} = \mathbf{a}_x QV/S$$

where M is the mass of the charged particle. The resulting equation of motion for the charged particle is

$$d^2x/dt^2 = QV/MS$$

The trajectory solution is subject to various boundary conditions.

In laboratory voltammetric determinations, background electrolytes are normally used, as indicated in Section 3.2, and migration effects are negligible (Adams, 1969).

3.3.2. Linear Diffusion

We assume as boundary conditions that the electrolytic cell is large relative to electrode size and that a background electrolyte is used; thermal equilibrium has been established and the system is mechanically quiescent. Thus migration and convection may be neglected. The basic reaction, as stated in Section 3.2 is

$$\text{oxidant} + ne \rightleftarrows \text{reductant}$$

Simple linear diffusion is defined by Fick's first law:

$$dN/dt = DA \, \partial C/\partial x$$

where

N = number of moles of substance involved

A = cross-sectional area through which the substance is diffusing

D = diffusion constant

C = concentration of substance

x = linear distance coordinate

It is sometimes convenient to write Fick's first law on a per-unit-area basis. Thus $dN/A\,dt$ is defined as flux ϕ (moles/sec/unit area), and Fick's first law is expressed as

$$\phi = D\,\partial C/\partial x$$

If the two electrodes in our cell are located at $x = 0$, $x \to \infty$, and we select two planes between them located at $x, x + \Delta x$, then the time rate of concentration change is

$$\Delta C/\Delta t = [\phi(x + \Delta x) - \phi(x)]/\Delta x$$

on a unit-area-flux basis. Passing to the limit,

$$\partial C/\partial t = \partial \phi/\partial x$$

and substituting from the first law,

$$\partial C/\partial t = D\,\partial^2 C/\partial x^2$$

which is Fick's second law of linear diffusion.

Adams (1969) has evaluated this equation for the oxidant (subscript o) for the boundary conditions

$$\text{for } t = 0, \qquad C_o^e = C_o^b$$

$$\text{for } t > 0, \qquad C_o^e = 0$$

$$C_o^e \to C_0^b \quad \text{as} \quad x \to \infty$$

where e and b refer to the electrode surface and the bulk solution respectively. The solution to the diffusion equation under these boundary conditions is

$$C_o(x, t) = C_o^b\,\mathrm{erf}(X)$$

where

$$X = x/(2\sqrt{Dt})$$

The error function $\mathrm{erf}(x)$ is evaluated from tables.

For the reductant,

$$C_r(x, t) = C_o^b\,\mathrm{erfc}(X)$$

where $\mathrm{erfc}(X) = 1 - \mathrm{erf}(X)$.

From these relations, the concentration–distance profiles can be computed for fixed values of time during the electrolysis process.

Actual charge flow is proportional to the flux at $x = 0$ and can be expressed by

$$dq/dt = nFA\phi(0, t)$$

which, upon substitution for $\phi(0, t)$, yields

$$dq/dt = nFAD \, \partial C_0(0, t)/\partial x$$

$\partial C_o(0, t)/\partial x$ can be evaluated by differentiating the error function relation. The charge flow in time, dq/dt, can be expressed as an electric current i which is a measurable quantity,

$$i = nFA\sqrt{D_o}C_o^b\Big/\sqrt{\pi t}$$

where

A = area, m^2

C_o^b = bulk concentration, moles/liter

D_o = diffusion coefficient, m^2/sec

F = Faraday's constant, C/mole

i = instantaneous current, A

n = number of electrons in total electrode reaction

t = time, sec

Theoretically, by measuring i, D_o can be determined for some time t, from this relation, if the other quantities are known.

Delahay (1954) has discussed cylindrical and spherical electrode geometries. In our notation, the instantaneous current is

$$i_c = nFAD_oC_o^b/r[(\pi\alpha)^{-\frac{1}{2}} + \tfrac{1}{2} - \tfrac{1}{4}\sqrt{\pi/\alpha} + \ldots]$$

$$\alpha = D_ot/r^2, \qquad r = \text{radius of cylinder}$$

$$i_s = 4\pi rnFD_oC_o^b + nFAC_o^b\sqrt{D_o/\pi t}$$

$$r = \text{radius of sphere}$$

3.3.3. Convection

Spontaneous convection occurs in any unstirred solution when electrolysis is present. Treatment is beyond this text. Theories have been proposed by Agar (1947), Levich (1944), and Wagner (1949).

Forced convection may be effected by stirring the solution, rotating or vibrating electrodes (or both), or working with a flowing solution. Theoretical treatment of forced convection in electrolytic cells is difficult, but is possible under certain assumed boundary conditions. A quantity of interest is the diffusion layer away from the electrode surface, and empirical relationships have been developed for some electrode geometries. Two relationships are summarized below:

Simple stirring:

$$d = \beta/u^n \qquad \text{Nernst relation} \qquad (n = \text{experimental constant})$$

Rotated disc electrode:

$$d \sim 3\sqrt{v/\omega}$$

where

$$d = \text{thickness of diffusion layer, m}$$

$$\beta = \text{experimental constant}$$

$$u = \text{velocity of liquid flow, m/sec}$$

$$v = \text{Maxwell kinematic viscosity, m}^2/\text{sec}$$

$$\omega = \text{angular velocity of electrode, rad/sec}$$

A detailed treatment of this subject, including limiting currents, is given by Adams (1969) and the associated references. Additional theoretical treatment of mass transport phenomena is given by Newman (1973).

3.4. Some Electrode Reactions

As shown in Section 3.2, the basic voltammetric reaction is

$$\text{oxidant} + ne \rightleftarrows \text{reductant}$$

This reaction has three components:

$$(\text{oxidant})_{\text{bulk}} \longrightarrow (\text{oxidant})_{\text{at electrode}}$$

$$(\text{oxidant})_{\text{at electrode}} + ne \longrightarrow (\text{reductant})_{\text{at electrode}}$$

$$(\text{reductant})_{\text{at electrode}} \longrightarrow (\text{reductant})_{\text{bulk}}$$

The first two of these components are reactions which proceed with finite rates and are mass-transfer controlled.

In an electrode–electrolyte system, oxidation occurs at the anode and reduction at the cathode (Skoog and West, 1963). When a reaction takes place at an electrode, a change in potential energy occurs; energy is either required for the reaction to proceed, or energy is released. At the standard conditions of 1 atm, with all dissolved substances in 1 molal concentrations and at a temperature of 298°K, the reaction potential is defined as $E°$.

Under other than standard conditions, the potential is estimated by the Nernst equation:

$$E = E° - (RT/nF)\ln Q$$

where

E = reduction potential for existing conditions

$E°$ = standard reduction potential

R = universal gas constant (8.314 V/mole/K°)

T = absolute temperature (°K)

n = number of electrons participating in reaction

F = Faraday constant (96,493 C/mole)

Q = ratio of the products of the activities (fugacities) of the reaction products to the products of the activities (fugacities) of the reactants, with each activity raised to a power equal to the exponent of that substance in the chemical reaction.

For the general reaction

$$\alpha A + \beta B + \cdots + ne \rightarrow \gamma G + \eta H + \cdots$$

Q is expressed as

$$Q = ([G]^{\gamma}[H]^{\eta}\ldots)/([A]^{\alpha}[B]^{\beta}\ldots)$$

where the brackets denote activity of the substance (Latimer, 1952).

As one example, let us examine a system studied by Grossman (1973). The cell uses two planar gold electrodes in aqueous NaCl solution (0.12N). Initially Na^+, Cl^-, H^+, and OH^- are present in solution. A steady current is passed through the cell from an external source.

The basic cathode reaction is

$$2H_2O + 2e \rightarrow 2OH^- + H_2 \uparrow$$

with a standard reduction potential for this reaction of -0.8277 V.

Three anode reactions are possible:

$$Au + 3Cl^- \rightarrow AuCl_3 + 3e^- \tag{1}$$

$$2H_2O \rightarrow 4H^+ + 4e^- + O_2 \uparrow \tag{2}$$

$$2Cl^- \rightarrow 2e^- + Cl_2 \uparrow \tag{3}$$

Adams (1969) has suggested that (1) occurs whenever gold is used as an electrode in a solution that contains chloride ions. Smith et al. (1964) have suggested (2), while Mason and Juda (1959) described the expected reaction (3).

The respective potentials of these reactions are:

(1) -1.00 V
(2) -1.229 V
(3) -1.358 V

The products of these reactions at the electrode surfaces may further react to yield other products. Possible additional reactions are:

$$Cl_2 + H_2O \rightarrow HCl + HOCl$$

$$2HOCl + NaCl \rightarrow NaOCl + Cl_2 + H_2O \qquad \text{(Jacobson, 1948)}$$

The chlorine gas remains dissolved in solution.

From analysis of this simple cell, we see that a rather complex chain of reactions can occur. When we add rate-sensitive mass-transfer phenomena and broad-spectrum currents (pulses), exact analysis becomes impossible, and we must resort to experimental observations of the sort described in Chapter 2 (see also Section 4.4).

Thus the potential associated with any electrode reaction is not only related to the reaction itself, but also to temperature, pressure, and the activities of the reactants and their products. As suggested previously, the processes which occur at an electrode surface are not necessarily the same as those which occur in the bulk electrolyte. In electrode reactions, the chemical activities involved are those at the electrode surface (Skoog and West, 1963), and these are virtually impossible to determine in most cases; hence, use of the Nernst equation to predict cell potentials is an approximation at best.

When we introduce rate-sensitive mass-transfer processes, analysis becomes complex if not impossible. Newman (1973) has stated: "... electrode kinetics are, in general, neither predictable nor reproducible on solid electrodes."

These considerations are the reason behind the phenomenological approach taken in Chapter 2 with respect to alternating-current electrode polarization effects. It is virtually impossible to determine rate constants and surface-chemical environments under the conditions of applied broad-spectrum signals.

3.5. References

Adams, R. N., 1969, *Electrochemistry at Solid Electrodes*, Marcel Dekker, New York.

Agar, J. N., 1947, *Discussions Faraday Soc.* **1**:26.

Delahay, P., 1954, *New Instrumental Methods in Electrochemistry*, John Wiley (Interscience), New York.

Grossman, R. S., 1973, *An Electrically Active Artificial Membrane System*, Ph.D. dissertation, University of Wyoming, Laramie.

Jacobson, C. A., 1948, *Encyclopedia of Chemical Reactions, Vol. 2*, Reinhold, New York.

Kolthoff, I. M. and Lingane, J. J., 1952, *Polarography, Vol. 1*, 2nd ed., John Wiley (Interscience), New York.

Latimer, W. M., 1952, *The Oxidation States of the Elements and Their Potentials in Aqueous Solutions*, 2nd ed., Prentice-Hall, New York.

Levich, B., 1944, *Acta Physicochem. USSR* **19**:117, 133.

Mason, E. A. and Juda, W., 1959, Applications of ion exchange membranes in electrodialysis, in *Adsorption, Dialysis, and Ion Exchange: Chem. Eng. Progr. Symp. Ser.* **55**(24):155–162.

Newman, J. S., 1973, *Electrochemical Systems*, Prentice-Hall, Englewood Cliffs.

Skoog, D. A. and West, D. M., 1963, *Fundamentals of Analytical Chemistry*, Holt, Rinehart, and Winston, New York.

Smith, A. L., Berkowitz, H. D., and Bluemle, L. W., Jr., 1964, Electrodialysis of blood: Evaluation of a high capacity unit, *Trans. Am. Soc. Artificial Internal Organs* **10**:273–279.

Wagner, C., 1949, *Trans. Electrochem. Soc.* **95**:161.

CHAPTER 4

Microelectrodes

Microelectrodes fall into two general categories, metal and glass pipette. They are distinguished from gross electrodes by the fact that active electrode surfaces are small enough to contact a single cell or neural unit. Electrode tip sizes generally lie in the $0.25-5\,\mu$ range. In shape, microelectrodes are usually tapered needles, since this configuration combines a small tip diameter with a reasonably large shaft diameter. The latter is necessary for mechanical positioning of the electrode and for making necessary electrical connections to the recording system or excitation source. Figure 4.1 illustrates the nomenclature associated with microelectrodes.

This chapter treats both metal and glass microelectrodes with regard to fabrication, mechanical and electrical characteristics, and general use. Requirements for headstage amplifiers used with these electrodes are discussed in Chapter 7.

4.1. Metal Microelectrodes

In terms of the mechanics of assembly, metal microelectrodes are the simplest to produce. They are simply metal wires or needles which have very small tips. All of the electrode, save the tip, is insulated with a suitable material as shown in Figure 4.2. The only mechanical problem associated

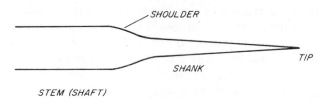

Figure 4.1. Microelectrode terminology.

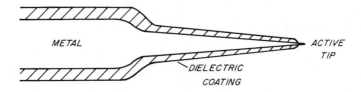

Figure 4.2. Metal microelectrode (cross-section).

with such electrodes is finding a stiff enough metal to insure electrode rigidity and suitable materials and techniques for providing the insulating sheath.

Electrodes can be made by firmly attaching a stiff wire to a shaft of larger diameter, or by drawing rod stock through a die. The shaft end is fitted to an appropriate electrical connector. The tip requires careful finishing before the electrode can be used. Normally, the tip is first ground smooth. Grinding rather than cutting is preferred as this prevents the tip from fraying. The final pointing of an electrode tip is done by electrolytic methods. The exact technique depends upon the electrode material.

4.1.1. Etching

Suitable metals for metallic microelectrodes are stainless steel, tungsten (tungsten carbide), and hard platinum (platinum–iridium alloy). The basic technique for electropointing is illustrated in Figure 4.3. A simple plating system is established with the electrode to be pointed as the anode. For stainless steel electrodes, Grundfest *et al.* (1950) have recommended the following electrolyte:

$$H_2SO_4 \qquad 34 \text{ ml (sp. gr. 1.84)}$$

$$H_3PO_4 \qquad 42 \text{ ml (sp. gr. 1.69)}$$

$$H_2O \qquad \text{enough to make 100 ml of electrolyte}$$

In this case, the cathode is a platinum rod. The battery voltage V (nominal 6 V), the resistance R, and the electrode separation S, are adjusted to give a cell current I of approximately 30 mA. (This is current per needle if more than one electrode is being prepared at the same time.)

To produce a tapered tip, 15–20 mm of the wire is immersed in the electrolyte and then slowly withdrawn. This step is repeated with the withdrawal speed adjusted to produce the desired degree of tapering. The needle must be regularly examined with a microscope. Later steps in the pointing should be done with reduced current to prevent erosion of the tip. When the desired degree of pointing has been achieved, the electrode is dipped in 10% HCl solution and washed in distilled H_2O and ethyl alcohol (95% C_2H_5OH).

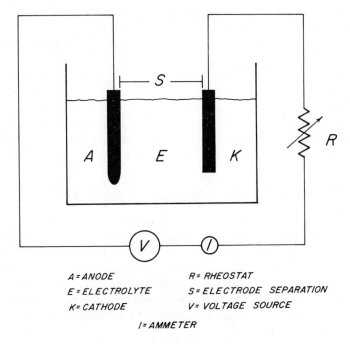

A = ANODE   R = RHEOSTAT
E = ELECTROLYTE   S = ELECTRODE SEPARATION
K = CATHODE   V = VOLTAGE SOURCE
            I = AMMETER

Figure 4.3. Electrode-pointing circuit.

This prevents staining of the metal. Techniques for applying insulation are discussed in Section 4.1.2.

A technique for pointing tungsten wire has been recommended by Hubel (1957). In Figure 4.3, K is a carbon rod, E is a saturated aqueous solution of KNO_2 or KOH, and V is 2–6 V ac. The tungsten wire, for convenience, can be held for pointing by passing it through a hypodermic needle and then crimping the needle. As before, the tip is produced by repeated insertion and withdrawal of the wire from the etching solution. By this process, it is possible to form tips less than 0.5 μ in diameter starting with 125 μ tungsten wire. After etching, the electrode is washed with detergent and coated with insulation.

4.1.2. Insulating

Various materials and techniques exist for applying the insulating sheath to metal microelectrodes. Some suitable materials are: Insl-X, E-33 lacquer (Insl-X Co., Inc. Ossining, N. Y.), clear vinyl lacquer 8-986S (Stover-Mudge, Inc., Pittsburgh, Pa.), Bakelite L3128 (Bakelite Corp.), Kem Lustral (Sherwin-Williams Co.), Formvar Enamel 9825 (General Electric Co.),

Glyptal (Walsco Electronics Mfg. Co.), 966 varnish (Dow Corning), and "Hypalon" P-6 (du Pont).

Dielectric coating is accomplished by dipping the needle into the coating material and slowly removing it, but generally one must experiment at first in order to develop a good technique for the particular combination of electrode and insulating materials being used. Usually several coats of thinned insulating material work best. The diluted coating material provides uniform insulation and will draw back from the tip to leave it exposed. Some coatings, such as Formvar, require baking (30 min at 120°C) after coating. Drying in a gentle flow of warm air is usually adequate for lacquers.

4.2. Metal-Filled Glass Micropipette Electrodes

One method for producing a well-insulated microelectrode is to start with a glass capillary tube and bond it to an internal metallic wire. Several techniques exist. A low-melting-point metal can be used to fill the tube, or a wire of the same diameter as the internal diameter of the tube can be passed through the glass capillary, which is then heated to produce bonding. One must be careful to select glasses and metals which have nearly the same temperature coefficients of expansion. Three basic techniques exist as described below.

4.2.1. Low-Melting-Point Glass and High-Melting-Point Metal

A somewhat crude but simple technique begins with a fine wire ; platinum, silver, or stainless steel can be used. Glass capillaries are drawn which have internal diameters just slightly larger than the wire diameter. The wire is passed through the glass tube (Figure 4.4) and a sharp flame is used to make the bead. The advantage of this type of electrode is ease of fabrication. There are several disadvantages. The final electrode diameter, tip excepted, is large. The bead area is fragile and subject to damage by wire flexion. The metal point may be damaged in the beading process. It is possible, although awkward, to etch the point after beading. A problem here is that the etching solution may erode or damage the glass bead. This is, however, a relatively simple way for producing a small silver–silver chloride electrode, as the tip can be chlorided after beading.

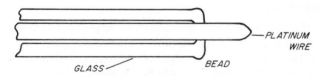

Figure 4.4. Glass bonded to metal electrode.

4.2.2. High-Melting-Point Glass and Low-Melting-Point Metal or Metal Alloy

An alternative technique is to use high-melting-point glass capillaries and low-melting-point metals such as indium (Dowben and Rose, 1953; Gesteland et al., 1959) or metal alloys such as silver solder (Svaetichin, 1951; Gray and Svaetichin, 1951).

The basic technique is as follows: Coarse (10–30 μ) capillary tubes are drawn out. The metal to be used is then melted into the tube lumen. The metal-filled tubes are then heated and drawn down to 1 μ or less (see Section 4.3 for pulling techniques). The electrode tips may then be plated, and they are usually platinized. Platinizing technique was discussed in Chapter 2.

Silver alloy electrodes made in this manner are usually rhodium-plated (tips) and then platinized. They serve well in registering fast transients and exhibit a noise figure close to the theoretical limit for their equivalent resistance.

When indium is used, a special glass-wetting alloy is required (Dowben and Rose, 1953). This alloy is composed of 50% indium and 50% Wood's metal (4 parts bismuth, 2 parts lead, 1 part tin, 1 part cadmium, melting point 71°C; alternative recipe: 50 parts by weight bismuth, 25 parts lead, 12.5 parts each of tin and cadmium, m.p. ~65°C). As in the previous case, coarse pipettes are drawn out and soaked for approximately 12 hr in dichromate cleaning solution for glass (35 ml saturated sodium dichromate (tech.) dissolved in 1 liter concentrated H_2SO_4; pour acid into dichromate solution). The tubes are then washed in distilled H_2O, flushed with ether, and dried. The metal alloy is heated to its melting point and the capillary tubes are prewarmed. After removing any oxide slag from the surface of the molten alloy, the bright metal is introduced into the tubes by suction. The alloy may be heated in an oil bath or on a hot plate. Once the tubes have been filled, they are heated again in a pipette puller, and micropipettes are pulled with 1.5–3 μ tips. It is necessary to keep the metal meniscus as close as possible to the pipette tip. The tip is filled with metal as follows: Insert a metal wire into the shaft end of the tube (the diameter of this wire should be approximately the lumen diameter; this wire will become the electrical connection to the microelectrode); put the assembly on a hot plate and push on the wire until the alloy nearly fills the tip of the pipette; remove the assembly from the hot plate and allow it to cool. After cooling, place just the tip end of the pipette on the hot plate so that metal flows out of the tip and forms a small ball; immediately remove the electrode from the hot plate and tap the shaft to knock off the ball. Use a microscope to check the electrode —the metal should be even with the tip and there should be no splattered metal on the glass; the tip region should be free from air bubbles. Before oxide can form on the metal, gold should be electroplated on the tip (Clark,

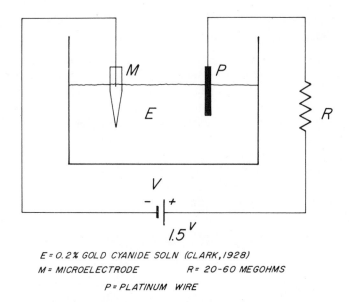

E = 0.2% GOLD CYANIDE SOLN (CLARK,1928)
M = MICROELECTRODE   R = 20-60 MEGOHMS
        P = PLATINUM WIRE

Figure 4.5. Circuit for tip-plating a glass–metal microelectrode.

1928). Figure 4.5 illustrates the plating cell. Plating is conducted for 30–60 sec. The electrode is then "prepoisoned" (Marmont, 1949) and platinized by the same technique as shown in Figure 4.5 with the following electrolyte: 100 ml 1% solution of chloroplatinic acid to which 0.01% lead acetate and 2 ml of cool, viscous gelatin solution have been added. This technique forms a 3- to 5-μ cap on the electrode tip, which is reported to enhance recording from single nerve cells.

4.2.3. Metal-through-Glass Microelectrodes

An alternative technique for glass-insulating platinum microelectrodes was developed by Wolbarsht *et al.* (1960). The electrodes are pushed through molten glass for coating. The technique is as follows: The metal electrode is produced by electropolishing 8–10 mil hard platinum wire (30% iridium, 70% platinum) in 30% NaOH saturated with NaCN. The cell (Figure 4.3) consists of a carbon cathode and a platinum wire anode. Initial polishing is done with $V = 6$ V dc and final polishing is done with $V = 0.8$ V rms, ac. Tips 1 μ in diameter can be obtained. The glass coating is produced by pushing the polished tip through a drop of molten soldering glass (Corning type 7570). The molten glass is supported on a hot U-shaped platinum filament. Microscope examination is necessary during the insulating process. Initially the glass is heated to a temperature above that required for adhesion of the glass to the platinum metal. Thus the electrode tip passes cleanly through the molten glass. Once the desired length of bare tip is seen extending through the glass, the filament temperature is reduced until the glass will stick to the platinum. The electrode is now pushed through the molten glass until the desired length of coating has been obtained. The rate at which the electrode is passed through the molten glass determines the insulation thickness. The tip may be platinized if desired. The configuration of Figure 4.5 is used. $E = 2\%$ $PtCl_2$ solution, $R = 1$ MΩ, $V = 15$ V dc.

4.3. Electrolyte-Filled Glass Microelectrodes

Many of the electrodes used in electrophysiological work are fluid-filled glass pipettes. The basic structure is illustrated in Figure 4.6. The electrode is a glass capillary drawn to a fine point. The lumen is filled with an electrolyte, usually an aqueous KCl solution, and a metal wire is inserted in the stem to form the electrical connection.

Of the three commonly available types of glasses, borosilicate (Pyrex-type) glasses appear to be the most satisfactory. They combine the properties of good electrical resistance, resistance to thermal shock, and mechanical

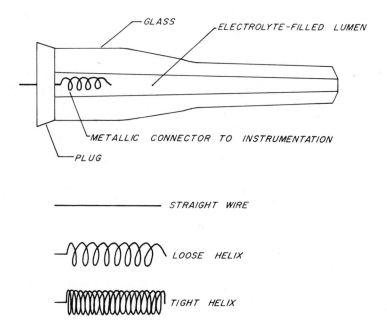

Figure 4.6. Glass microelectrode. Several configurations for the internal pickup electrode are shown.

strength. Soda-lime glasses have poor heat resistance and chemical stability, while lead glasses tend to be fragile in electrode work and also possess a high temperature coefficient of expansion which makes them subject to thermal shock.

A suitable glass for micropipette applications is Corning 7740 (Frank and Becker, 1964). Tubing with the dimensions 1.0 mm o.d. and 0.5 mm i.d. is satisfactory in most applications. Stock tubing should be selected to the tolerances of ± 0.1 mm for the outside diameter and ± 0.05 mm for the wall thickness to insure uniformity and reproducibility in pulling electrodes.

4.3.1. Electrode Pulling

Many types of electrode pullers are commercially available and many workers construct their own units for particular applications. A good summary of types of pullers is given by Frank and Becker (1964). Figure 4.7 illustrates the basic device. Both vertical and horizontal pullers are available. A vertical assembly is shown in Figure 4.7. The device has a rigid vertical support and base. A fixed clamp is located at the top to hold the glass capillary tubing firmly. The tubing is positioned through a heater element and a mass is clamped to the bottom of the tube. The heater temperature and the mass must be adjusted experimentally for the type of glass used. With proper adjustment of these parameters, the capillary tube will neck down and separate cleanly leaving an open tip $0.5–1 \mu$ in diameter. A simple mass acting under the influence of gravity can be used. More sophisticated devices use a spring and solenoid arrangement in place of a mass. In this case, a very lightweight clamp is used on the bottom of the tube. Adjustable tension springs and variable-pull solenoids are used to achieve optimum pulling conditions. For glass of a given formula, the two variable system parameters are heater (filament) temperature and spring tension (effective mass). The heaters used are generally platinum foil or filaments. Temperature control is obtained by operating the filament from a low-voltage, high-current transformer (such as one used for spot welding), which in turn is controlled by an autotransformer (Variac). Reproducibility of micropipettes depends upon constancy of pulling tension, heater temperature, and heater position. The filament or foil must be secured rigidly to prevent sagging. If the filament sags during pulling, uneven electrodes may result, or too long a length of glass may be heated, producing an undesirably long shoulder on the electrode. Open wire filaments are the most susceptible to sagging. It is better to support wire filaments rigidly in a tube of refractory material such as Transite.

With proper cushioning of the mass or solenoid assembly after fracture of the glass has occurred, it is possible to produce two electrodes simultaneously, one in the fixed clamp and the other attached to the mass or solenoid unit.

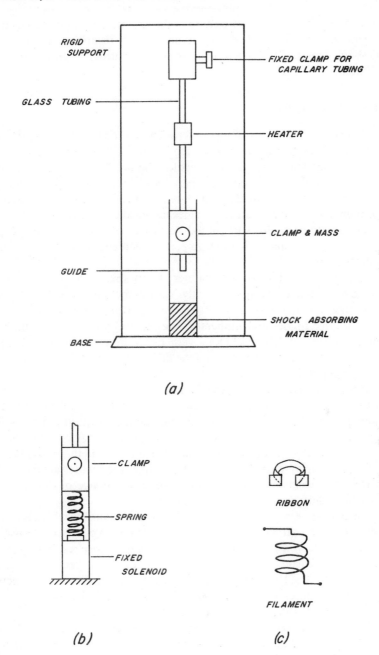

Figure 4.7(a). Simple microelectrode puller, (b) alternative "mass" spring and solenoid system, (c) two types of heater assemblies.

The length of the heater and, to some extent, the pulling force control the geometry of the finished micropipettes with regard to shoulder taper and shank length. Detailed discussion and performance graphs are given by Frank and Becker (1964).

4.3.2. Pipette Filling

A problem exists in filling micropipettes, especially those with tip diameters less than $5\,\mu$. Larger sizes can be filled with electrolyte introduced from a syringe via a hypodermic needle. The needle is pressed against the glass inside wall of the pipette's tapered section. An alternative technique is to use another micropipette in place of the hypodermic needle. A unit with approximately $5\,\mu$ tip diameter is nested inside the pipette to be filled. Filling operations of this sort should be carried out under a microscope to insure complete filling of the micropipette lumen and to prevent damage to the pipette tip.

Individual handling in this manner of very small pipettes is not recommended as tip damage occurs too easily. Various methods have been proposed for automatic filling. Basically these involve putting one or several pipettes into a suitable holder and immersing the pipettes and holder in the filling electrolyte, which is then boiled for several hours. Boiling time and subsequent tip erosion damage can be reduced if the filling solution is first heated and the assembly placed in a vacuum.

An alternative method (Tasaki *et al.*, 1954) involves immersing the pipettes into 40°C methanol and then placing the assembly in a vacuum chamber. The chamber is evacuated and the alcohol allowed to boil for approximately 8 min. In this time the micropipettes are filled. The assembly is then removed from the vacuum chamber and the methanol replaced by the filling electrolyte. In about two day's time, the methyl alcohol will be replaced with electrolyte by diffusion. The disadvantage of this process is the two-day delay.

Ideally the filling electrolyte should be a saline solution which is isotonic to the physiological system in which the microelectrode is to be used. In practice, with such solutions, too high electrode resistance results (see Section 4.4). For this reason 3M KCl is frequently used. Other electrolytes include 1% NaCl, 2M NaCl, 0.6M K_2SO_4, and others. In some instances, electrode application dictates what electrolyte should be used (Eccles *et al.*, 1957; Boistel and Fatt, 1958; Ito *et al.*, 1962).

To prevent electrode plugging, freshly prepared and filtered electrolytes should be used. They should be boiled to sterilize them to avoid bacterial growth in the micropipette lumen and contamination of the life system being studied.

Generally, micropipette electrodes should be stored in distilled water or alcohol. If stored in the filling electrolyte, bacterial growth may occur, and strong electrolytes tend to erode fine tips. Storage in the dark at low temperature retards bacterial growth. Micropipettes stored in and filled with distilled water or alcohol can be filled with electrolyte for use by the diffusion technique mentioned above. Electrolyte-filled electrodes stored in air are frequently damaged by crystallization of salts on their tips, with resultant tip fracture.

4.4. Electrical Properties of Microelectrodes

When a metal is placed in contact with an electrolyte, a potential difference is observed at the liquid–metal interface, as noted in Chapter 2. This is similar to the work-function potential difference which occurs when two dissimilar metals are brought into contact, or the potential difference associated with a semiconductor p–n junction. The value of potential difference associated with a metal electrode–electrolyte interface is a function of the metal and contacting electrolyte. Theoretical treatment of this situation is complex and one should refer to a text on electrochemistry such as those by MacInnes (1961) or Newman (1973). Certain types of electrodes are extremely sensitive to various trace impurities in the contacting electrolyte and may react quite differently in seemingly similar circumstances.

When current is passed through paired electrodes in contact with an electrolyte, various reactions occur. In biological work, one normally deals with an "active" electrode and an "indifferent" or reference electrode. A typical microelectrode system is shown in Figure 4.8. In two-terminal impedance measurements as discussed in Chapter 2, two "active" electrodes are involved.

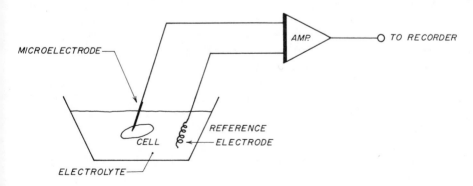

Figure 4.8. Basic microelectrode recording system.

During the passage of electric current in an electrolyte, the various ions in solution carry different proportions of the current. The numerical value of each fraction is called the *transference number* of the associated ion. Techniques for determining these numbers were worked out by J. W. Hittorf (1853, 1856, 1858, 1859). In a simple electrode–electrolyte system of two silver electrodes in contact with silver nitrate solution, the following reactions occur with passage of current: Per faraday passed, the electrode reaction at the anode is

$$Ag = Ag^+ + e^-$$

and the cathode reaction is

$$Ag^+ + e^- = Ag$$

Between the electrodes, current is carried partly by the silver ions and partly by the negatively charged nitrate ions. This leads to a situation which is analogous to space charge formation in a thermionic vacuum tube. Since the Ag^+ ions are only partly responsible for the current in the electrolyte, they are not transported away from the anode immediately upon formation. There is, then, an accumulation of silver ions about the anode and a depletion of silver ions at the cathode. These phenomena can be observed in a Hittorf transference apparatus. Similar phenomena occur in mixed systems, for example, platinum electrodes in contact with sodium chloride solution.

The net effect of these phenomena is the buildup of a charge (metal ion) concentration gradient in the solution. Cessation of current, or reversal in polarity, gives rise to a diffusion potential which is opposite in sense to the current which generated it. When the polarity of two electrodes is reversed, there is a time delay in buildup of positive metal ions at the anode (formerly the cathode, before reversal) and the depletion region about the cathode (formerly the anode). With alternating currents or biphasic pulses, current reversal may occur quite rapidly. As the frequency of an alternating current is increased, less ionic motion occurs (because of the time delay just noted), and less change occurs in the charge concentration at the metal–electrolyte interfaces. As frequency is further increased, electric charge transfer through the electrolyte approaches that in a dielectric; that is, an electric dipole (double) layer is produced at the metal–fluid interface. The interface then behaves as a capacitor. This gives rise to the variation of impedance inversely with frequency as noted in gross electrodes (Chapter 2). The situation in microelectrodes has been reported by Gesteland *et al.* (1959). This is the phenomenon described in Chapter 2 as alternating-current electrode polarization. The situation with microelectrodes is generally more complex as direct current and pulsed signals are frequently involved, rather than simple sinusoidal signals.

Whether metal or fluid-filled microelectrodes are used, the interface effect is still felt. In the case of metallic electrodes, the interface occurs at the "active" surface of the electrode. With fluid-filled micropipettes, the metal–electrolyte interface exists in the pipette stem where electrical connection is made to the recording or stimulating electronics. There is the additional complication of a two-electrolyte interface at the electrode tip, since the electrolyte filling the lumen is usually different from the external electrolyte.

For platinum electrodes in aqueous NaCl solution, we have at the anode

$$\tfrac{1}{2} HOH - e^- = \tfrac{1}{4} O_2 \uparrow + H^+ \qquad \text{(dilute solution)}$$

$$Cl^- - e^- = \tfrac{1}{2} Cl_2 \uparrow \qquad \text{(concentrated solution)}$$

and at the cathode

$$HOH + e^- = \tfrac{1}{2} H_2 \uparrow + OH^- \qquad \text{(dilute solution)}$$

$$HOH + e^- = \tfrac{1}{2} H_2 \uparrow + OH^- \qquad \text{(concentrated solution)}$$

The upshot of this discussion is that the impedance of microelectrodes is high for direct or low-frequency currents and decreases as frequency increases. Metal–metal chloride or reversible electrodes will be considered in a subsequent chapter.

4.4.1. Resistance of Microelectrodes

The resistance of a conductor of uniform right cylindrical geometry is given by

$$R = l/\kappa A \ \Omega$$

where

$$l = \text{length of cylinder, cm}$$

$$A = \text{cross-sectional area, cm}^2$$

$$\kappa = \text{conductivity, mhos/cm} \qquad \text{(metal or electrolyte)}$$

This approximates the resistance of the uniform stem-section lumen of a micropipette. The tip section of a microelectrode is approximated geometrically as a cone truncated by two concentric spheres (Amatniek, 1958) as shown in Figure 4.9. The resistance of a microelectrode of this configuration as measured between the two ends is

$$R = \frac{l\phi^2}{2\pi d'[1 - \cos(\phi/z)](d' + l\phi)\kappa}$$

These relationships are given in Figure 4.9. When the tip taper angle ϕ is

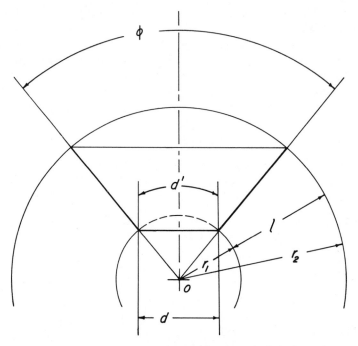

Figure 4.9. Geometric representation of microelectrode tip (redrawn from Amatniek, 1958, Figure 18, p. 14). O, origin and apex of concentric spheres and cone; r_1, r_2, respective radii of internal and external spheres; l, distance between the surfaces of the two spheres and the length of the truncated cone; d, diameter of circle formed by intersection of cone and smaller sphere; d', projection of d onto smaller sphere; ϕ, cone angle in radians.

small and when the tip diameter $d \ll l\phi$, the internal resistance of the electrode becomes

$$R = 4/(\pi\phi d\kappa)$$

Amatniek has also shown that the resistance between the electrode tip and a large external electrode (such as the indifferent electrode) is given by

$$R' = (\pi d\kappa')^{-1}$$

where κ' is the conductivity of the external medium.

4.4.2. Equivalent Circuit for a Microelectrode

Figure 4.10 illustrates a simple intercellular microelectrode recording situation. Figure 4.10b depicts the system with the microelectrode grossly enlarged to accommodate clarifying detail. Figure 4.10c is the electrical

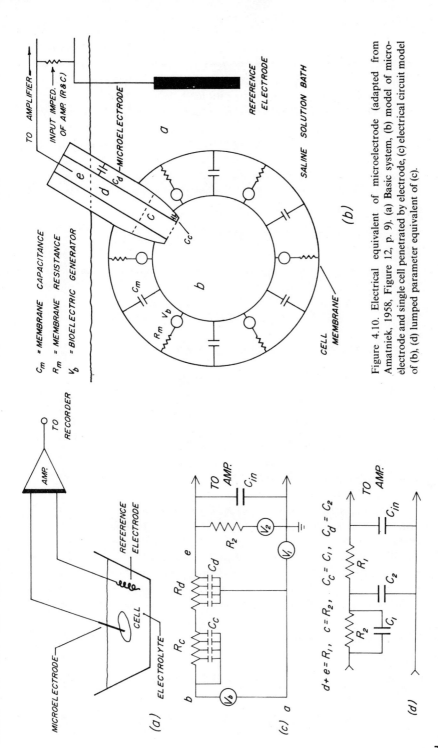

Figure 4.10. Electrical equivalent of microelectrode (adapted from Amatniek, 1958, Figure 12, p. 9). (a) Basic system, (b) model of micro-electrode and single cell penetrated by electrode, (c) electrical circuit model of (b), (d) lumped parameter equivalent of (c).

71

equivalent of 4.10b. An approximate lumped-parameter representation is shown in Figure 4.10d. C_1 is the capacitance between the cellular fluid or external electrolyte and the electrode tip region. R_1 is the resistance of the electrode tip region. C_2 is the capacitance between the electrode shaft and an external electrolyte. R_2 is the resistance of the shaft lumen. Any additional stray capacitances will be lumped with the cable capacitance of the connection to the external electronics (see Section 4.5 and Chapter 7 on preamplifiers). C_1 and C_2 are distributed capacitances. Further discussion of this subject appears in Sections 4.5 and 4.6.

4.4.3. Electrical Noise in Micropipettes

Metallic microelectrodes have the advantage over glass micropipettes that they produce lower amounts of electrical noise. Gesteland *et al.* (1959) have shown that noise in metal microelectrodes is approximately that for a pure resistance; that is,

$$v_{rms} = (4kTR\,\Delta f)^{\frac{1}{2}}$$

where

k = Boltzmann's constant

T = temperature of electrode, °K

R = real part of the electrode impedance, Ω

Δf = frequency band passed by the electrode, Hz

For a typical metal microelectrode, the rms (root-mean-square) noise voltage for the audio band of frequencies is of the order of 60 μV for a 10-MΩ electrode at room temperature.

Fluid-filled glass micropipettes are inherently more noisy than their metal counterparts. In addition to "$kTR\Delta f$" noise associated with the metal–electrolyte connector in the stem, we must deal with ionic flow. Ion flow occurs between the electrolyte in the lumen and the external electrolyte of the medium in which the electrode is placed. Ion movement depends upon tip size, ion concentrations in the two electrolytes, current in the electrode, pressure differentials, and other factors. It is a complex problem and not amenable to easy theoretical treatment. In general, the smaller the tip diameter, the lower the noise figure for glass electrodes. Little ionic flow occurs in pipettes with tips $<1\ \mu$ diameter.

4.5. Alternating-Current Electrode Polarization in Microelectrode Systems

Figure 4.10a illustrates schematically a rudimentary microelectrode recording system and Figure 4.11 its electrical equivalent. Both R_p and C_p

are functions of frequency and change value for each spectral component associated with the source potential V_s. As a consequence of the variable time constant $R_p C_p$ and the fixed time constant $R_i C_i$, the voltage V_r presented to the input of the recording amplifier may be quite different from the actual voltage waveform produced by the biological source. C_s and R_e are not present in all cases. C_s represents the capacitance between a metal microelectrode and the potential-generating surface and occurs in extracellular recording. It also appears (shunted by a resistance) in glass microelectrode systems and represents a capacitance between the metal contact within the

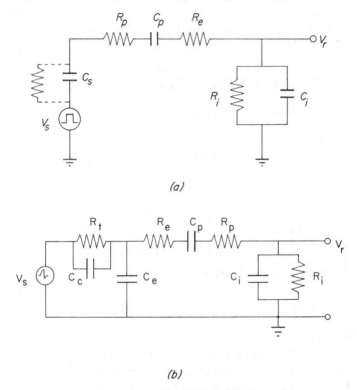

(a)

(b)

Figure 4.11. Two lumped-parameter circuit models for studying the effects of ac electrode polarization in microelectrode systems. V_s, simulated biological signal source; C_s, C_e, contact or source capacitances; C_p, electrode polarization capacitance $= f(\omega)$; C_e, shunt capacitance of electrode to electrolyte and reference electrode; C_i, cable leakage capacitance and input capacitance to amplifier system; R_e, resistance of microelectrode; R_p, electrode polarization resistance $= f(\omega)$; R_i, input resistance of amplifier system; V_r, input voltage to recording amplifier; ω, radian frequency ($2\pi f$).

body of the electrode and the external recording site. R_e is the series resistance in a glass microelectrode or the surface resistance between stimulating and recording electrodes in an extracellular multiple-electrode system. In the circuit model, the series C elements and the shunt R produce a high-pass filter or differentiator configuration, while the series R elements and the shunt C produce a low-pass filter or integrator configuration. Taking these combinations together, the overall configuration is a band-pass filter or integrator-differentiator, but with frequency-varying parameters, as R_p and C_p are frequency dependent. As a consequence of the electrical properties of this circuit, the output signal $V_r(t)$ may be substantially different from the source potential $V_s(t)$.

One can write a Laplace transform system transfer function for the circuit model. This is (neglecting the resistor shunting C_s)

$$G(s) = \frac{sR_iC'_p}{s^2R_i(R_e + R_p)C_iC'_p + s[R_iC_i + (R_e + R_p + R_i)C'_p] + 1}$$

$$= V_r(s)/V_s(s) \qquad \text{(applies to Figure 4.11a)}$$

where

$$V_r(s) = \mathcal{L}\{V_r(t)\}$$

$$V_s(s) = \mathcal{L}\{V_s(t)\}$$

where

$$C'_p = C_p \qquad\qquad \text{if } C_s \text{ is not present}$$

$$C'_p = C_pC_s/(C_p + C_s) \qquad \text{if } C_s \text{ is present}$$

For a single metal microelectrode, one may assume $R_e = 0$, and

$$G(s) = \frac{sR_iC'_p}{s^2R_iR_pC_iC'_p + s[R_iC_i + (R_p + R_i)C'_p] + 1}$$

Theoretically, it is now possible to calculate the inverse problem. Experimentally, $V_r(t)$ can be determined as this is the recorded signal, usually an action potential. With some effort, $V_r(t)$ can be specified as a function of time and its Laplace transform computed. The Laplace transform $V_s(s)$ of the input signal (action potential, etc.) $V_s(t)$ is computed from the relations

$$V_s(s) = V_r(s)/G(s)$$

$$V_s(t) = \mathcal{L}^{-1}\{V_s(s)\}$$

These relations hold provided that C'_p, C_i, R_e, R_i, and R_p are known. Herein lies the difficulty. C_i, R_e, and R_i are system parameters which are easily determined, but C'_p and R_p are functions of frequency, hence functions of the complex frequency s, and can only be approximated through the Fricke gamma-function relations. The C_s component of C'_p cannot be estimated easily. It would be possible to establish a computer program to perform the inverse problem provided that values were assumed for C_p and R_p as functions of frequency. It would be necessary, however, to perform one calculation for each spectral component of $V_r(s)$ and to assume a different value for C_p and R_p for each of these spectral components. The assumed values for C_p and R_p would have to be consistent with a known frequency behavior for electrode polarization. This would mean extensive experimental work on a given electrode system in order to determine these parameters.

Thus although theoretical calculations are possible, they are not practical, as there are undetermined and difficult-to-determine system parameters. For this reason, we turn now to a purely experimental approach.

4.5.1. Experimental Results

An experimental circuit in the configuration of Figure 4.11 was constructed. The electrode polarization impedance was simulated by using actual electrodes in a modified electrolytic cell configuration. Microelectrodes were placed in opposition to a large, flat, platinized platinum plate. By this means, the major current density is associated with the microelectrode rather than with the plate, which was several square centimeters in surface area. Physiological saline was used as the electrolyte. An electronic pulse generator was used to replicate the signal source $V_s(t)$. One-millisecond pulses were used to be consistent with the range of many experimentally observed action potentials. It was experimentally determined that the electrode polarization impedance of this model system was in the linear range for input signal intensities varying from 10 V peak to the limit of the oscilloscope resolution (< 10 mV) (see Ferris and Stewart, 1974).

The model circuit was constructed so that C_s, C_i, R_e, and R_i could be varied. The values which were chosen were consistent with typical microelectrode systems. To simulate glass microelectrodes, R_e was selected in the range from 10^7 to 10^8 Ω. C_i was varied from 0 to 100 pF, which is consistent with the summation of cable and input capacitances in a typical experimental setup. C_s was varied from 10 pF to infinity (direct contact with the signal source). The recording amplifier was a negative input capacitance unit (modified Bak system) designed for electrophysiological recording. A Tektronix type 564 storage oscilloscope (multichannel input) and camera provided the recording system.

The signal source pulse $V_s(t)$ was simply a rectangular pulse of 1-msec duration. Pulse amplitude could be varied from 0 to 10 V peak without overloading the amplifier system or affecting the linear range of the electrode polarization impedance. In each of the photographic records which follow, the signal source pulse is shown to provide time-base calibration. In most of the records, the output signal $V_r(t)$ is on a different amplitude scale from the input signal. For recording purposes, the modified Bak amplifier system was adjusted for unity gain.

4.5.2. Photographic Records

In Figure 4.12a, C_s is replaced by a short circuit and C_i is varied over the range specified above. Input capacitance compensation is not provided. The circuit configuration of Figure 4.12b is the same as that for Figure 4.12a, except that the amplifier has been adjusted to provide input capacitance compensation. One can compensate for input capacitance only within the range of the input amplifier. In Figure 4.12c, $C_i = 0$ and C_s is varied over the range from 10 pF to infinity. No input capacitance compensation is provided (this simulates the input conditions for a normal high-gain, high-input-impedance amplifier with some internal shunt capacitance). Figure 4.12d is the same circuit as used for 4.12b, except that input capacitance compensation is provided. Exact replication of the input signal $V_s(t)$ is provided only in one case, that is, when C_s is infinite (the direct contact case). Figure 4.12e represents an attempt to compensate for system capacitance, and overcompensation has occurred with ringing or oscillation in the amplifier feedback system, as shown by the spikes along the trailing edge of the signal. Here it was attempted to compensate for C_s. Another example of overcompensation of capacitance is shown in Figure 4.12f. In this case C_i is involved. Figures 4.12g, h, and i are typical waveforms with selected values of C_i and C_s. Figure 4.12j is the waveform produced when compensation is provided for C_i, $R_e = 0$, and C_s is varied. This represents a situation which can occur with extracellular metallic microelectrodes.

The records represented by Figures 4.12g, h, and i are very close to action potentials which are recorded in the laboratory and reported widely in electrophysiological literature. It is interesting to note that the experimental records figured are purely a result of electrode recording system artifact when a rectangular input pulse signal $V_s(t)$ was employed. On this basis, we can only infer that serious distortion of the true biological signal $V_s(t)$ can occur in standard electrode systems. Negative input capacitance compensation provided by some amplifiers designed for electrophysiological recording can provide signal correction only in certain instances, as indicated by the figures.

Because of the frequency behavior of $\dot{Z}_p$, it is difficult to predict just how serious signal distortion may be in a given recording system. For multielectrode systems such as are frequently used in neural stimulation and recording, a suggestion is made. Researchers frequently go to extreme lengths to eliminate the "artifact" signal in their recordings. The artifact signal which

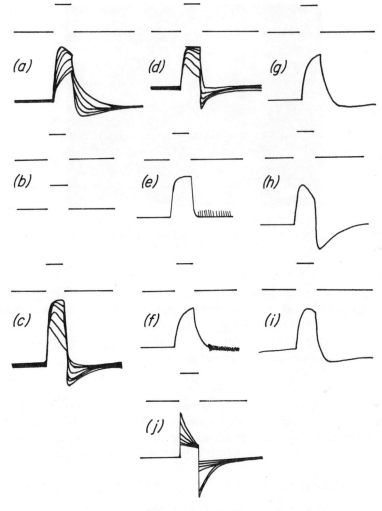

Figure 4.12. Ink tracing of oscillograph records from simulated microelectrode system. The original photographs are presented by Ferris (1971).

represents ohmic conduction along the surface of an axon can be extremely useful. If the recording system time base is expanded to portray the artifact clearly, the artifact signal can be compared to the stimulating signal, also expanded to the same time base. If the two signals, artifact and stimulus, have the same profile (there may be amplitude attenuation in the artifact relative to the stimulus), then no distortion can be assumed. On the other hand, if distortion does occur, its extent can be estimated (see Figure 4.13).

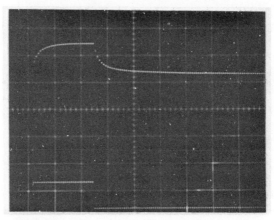

Figure 4.13. Bottom trace: stimulus signal applied to
frog sciatic nerve. Top trace: artifact signal at recording
electrodes. Note distortion. Calibration: 0.2 V/cm,
50 μsec/cm.

Let us represent the stimulus as $V_s(t)$ and the artifact as $V_a(t)$. If one
wishes to go to the effort, these can be represented by mathematical expressions. One can then find the Laplace transforms $V_s(s)$ and $V_a(s)$ which cor-
respond to the time functions $V_s(t)$ and $V_a(t)$, respectively. An electrode system
transfer function can now be defined as

$$G(s) = V_a(s)/V_s(s)$$

Now if the same stimulus $V_s(t)$ produces an action potential $V_r(t)$, we can
characterize this action potential mathematically and estimate a correction
for system distortion. Suppose that the action potential is represented
mathematically as a function of time $V_r(t)$ and we find its equivalent transform
$V_r(s)$.

The estimated corrected function is found by the following operations:

$$V_r'(s) = V_r(s)/G(s)$$

where $V_r'(s)$ is the Laplace transform of the "corrected" action potential.
Now

$$V_r'(t) = \mathscr{L}^{-1}\{V_r(s)\}$$

where $V_r'(t)$ is now the estimated "true" action potential. In reality, for the
artifact signal, a $\dot{Z}_p$ is present at both the stimulating and recording sites.
For the action potential, a $\dot{Z}_p$ is present at the recording site only. Thus
$V_r'(t)$ is an overcorrected signal, and only an estimated correction is possible
at best.

Generally speaking the distortion introduced by electrode polarization impedance will be higher at the stimulating electrodes than at the recording electrodes, as higher current densities are involved during stimulation. Thus the artifact signal will undergo greater distortion than the recorded action potential.

4.6. Microelectrodes—A Few Final Notes

The series resistance of a glass microelectrode is generally quite high. If isotonic saline solution (0.15N) is used as the electrolyte, a typical glass electrode will exhibit a series resistance of the order of 1000 MΩ. Such a resistance would cause too high a voltage drop with respect to the usual input impedance values of microelectrode preamplifiers (see Chapter 7 on preamplifiers). For this reason, 2 or 3N KCl solution is used to fill the electrode lumen. This results in an electrode series resistance of 50 to 100 MΩ.

In connection with associated electronic circuitry, fluid-filled microelectrodes behave as low-pass filters, while metal microelectrodes act as high-pass filters (Gesteland *et al.*, 1959). The reasons for this are simply that glass electrodes exhibit high shunt capacitance and series resistance (Figure 4.10) while electrode polarization impedance associated with metal microelectrodes produces a frequency-dependent *RC* combination whose series impedance decreases as frequency increases.

Generally, metal microelectrodes are useful for stimulation, or for recording high-frequency processes. Glass electrodes are useful in recording from cells in which membrane processes are of interest. Unfortunately, there is no general-purpose microelectrode. Preamplifiers will be discussed in subsequent chapters, as other aspects of microelectrode use are considered.

4.6.1. Noise in Glass Microelectrodes

A recent study by DeFelice and Firth (1971) has shown that the electrical noise present in glass microelectrodes is in excess of the Johnson noise predicted by the Nyquist formula given in Section 4.4.3. If the microelectrode lumen is filled with an electrolyte of concentration n_2 and the external electrolyte in which the electrode is immersed has the concentration n_1, when $n_1 = n_2$ a noise voltage is observed as predicted by the Nyquist formula. When $n_1 \neq n_2$, a dc offset voltage is noted (tip potential) because of the electrolyte concentration gradient, and the noise voltage far exceeds that for Johnson noise.

DeFelice and Firth found that the mean square voltage fluctuations per cycle ($n_1 \neq n_2$) can be represented by

$$v^2 = 4kTR \, \Delta f + bh(f)$$

for the limited-bandwidth case. The factor $bh(f)$ is not separable and is determined from experiments. Electrode resistance is represented by

$$R = \frac{S \log(n_2/n_1)}{n_2 - n_1}$$

where S is a geometry-related factor for glass electrodes.

At present, the origin of excess noise when $n_1 \neq n_2$ is not understood.

4.6.2. Electrical Connections to Microelectrodes

The electrical connection from a glass microelectrode to its accompanying electronics is made by a metal wire inserted in the stem end of the lumen and in contact with the filling electrolyte. Early literature frequently refers to tungsten wire for this purpose, but there seems to be no valid basis for using this metal. Apparently it was available and had been used in place of antimony in certain pH electrodes. Because of its metallurgical properties, tungsten is difficult to form, and it has undesirable electrical characteristics.

Presently platinum or silver wire is generally used to connect into glass electrodes. In some cases, the silver wire is chlorided to make a reversible-electrode connector. Reversible electrodes are discussed in Chapter 5.

4.6.3. Polarization Phenomena

In using microelectrodes and near-microelectrodes, it is sometimes difficult to separate stimulating and recording electrodes with respect to electrode polarization phenomena. An example is the five-electrode system frequently used in single-neuron studies. Similar electrodes are used for both excitation and recording. The system is a closed electrical circuit, so that charge flow exists at both paired electrodes as well as at the grounding electrode. In the pulsed (broad-spectrum signal) case or under sinusoidal excitation, any current density at an electrode interface establishes an electrode polarization impedance as discussed in Chapter 2. This occurs both at stimulating and recording electrodes as a function of current density.

In the stimulating-electrode case, the polarization impedance produces a voltage drop at the interface between the electrode and the biological preparation. This can lead to errors in determining voltage thresholds in voltage-sensitive systems. For this reason, threshold measurements should be related, where possible, to current levels rather than voltage levels.

At a recording microelectrode, ac electrode polarization, in addition to producing a frequency-sensitive voltage drop, may also introduce signal distortion.

If direct-current signals or monophasic pulses of a chronic repetitive nature are involved, electrode corrosion and other interface phenomena may occur, as described in Chapter 3. In this instance, excitation electrodes will be more susceptible to damage than recording electrodes. Normally the current density in recording microelectrodes should be very small if not negligible.

4.6.4. Multiple Microelectrodes

A double-barreled glass microelectrode, shown in Figure 4.14, has been developed by Coombs *et al.* (1955). Such an electrode can be used for voltage-clamp (Section 7.8) experiments involving single cells. Current can be passed through one barrel to a remote electrode, while the resulting cell potential is measured at the other electrode. In a double-barreled pair, one electrode may be used in the stimulating mode while the other is used in the recording mode for stimulus–response measurements. Rush *et al.* (1968) have presented a detailed analysis of the electrical properties of double-barreled micro-electrodes.

Multibarreled microelectrodes, with as many as five barrels, have been used for altering the chemical environment of single cells, in drug studies, and in microelectrophoresis (Frank and Becker, 1964; Curtis, 1964). Coaxial glass microelectrodes have been described by Freygang and Frank (1959).

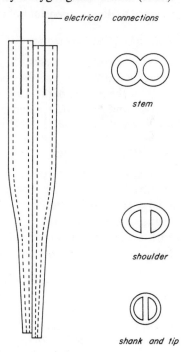

Figure 4.14. Double-barreled microelectrode.

Multibarreled electrodes are made by fusing together the requisite number of glass capillary tubes. The assembly is usually pulled in two steps: first, using a mass under gravitational influence and second, using a variable-pull solenoid assembly. A heavier-than-normal electrode puller is required for multibarreled electrodes. The tubes at the top end of the assembly are usually bent at various angles to one another to facilitate making the electrical connections.

Filling and storage of multibarreled micropipettes is a bit of a problem as each barrel may contain a different solution. Frank and Becker (1964) have suggested the use of filler pipettes. Another technique involves filling all of the barrels with distilled water using conventional filling methods. Most of the water is then removed by suction from each barrel in succession, and replaced by the filling solution. By this process, the tip initially remains filled with distilled water and the filling solution then diffuses into the tip region (Curtis, 1964).

Multibarreled electrodes are usually stored vertically with the tips immersed in distilled water, unless all of the barrels are filled with the same solution. In that case, the assembly may be stored in the filling solution. It is necessary to devise a special holder to maintain the vertical storage position.

4.6.5. pH Microelectrodes

pH electrodes are described in detail in Chapter 5. A microelectrode for measuring intracellular pH has been developed by Caldwell (1958). pH glass is used for the tip end of the capillary. The capillary is pulled to form a microelectrode and then the tip end is sealed by placing it near a hot platinum filament. A glass-membrane pH microelectrode is the result.

4.6.6. Pore Electrodes

It is sometimes necessary to isolate a glass electrode tip from a surrounding fluid. For this purpose pore electrodes, which have a small lumen but large tip cross section, are used. A thick-walled pipette is drawn and the end fused. By subsequent grinding of the end, a pore electrode is formed (Pratt, 1917). The final electrode possesses a small end opening (pore) with a large coaxial insulating-ring surface.

4.7. References

Amatniek, E., 1958, Measurement of bioelectric potentials with microelectrodes and neutralized input capacity amplifiers, *Trans. IRE, PGME* **10**:3–14.
Boistel, J. and Fatt, P., 1958, *J. Physiol.* (*London*) **112**:9P.
Caldwell, P. C., 1958, *J. Physiol.* (*London*) **142**:22.
Clark, W. N., 1928, *The Determination of Hydrogen Ions*, Williams and Wilkins, Baltimore.

Coombs, J. S., Eccles, J. C., and Fatt, P., 1955, The electrical properties of the motoneurone membrane, *J. Physiol.* **130**:291–325.

Curtis, D. R., 1964, Microelectrophoresis, in *Physical Techniques in Biological Research*, Vol. 5 (W. L. Nastuk, ed.), Academic Press, New York.

DeFelice, L. F. and Firth, D. R., 1971, Spontaneous voltage fluctuations in glass microelectrodes, *Trans. IEEE, BME* **18**:339–351.

Dowben, R. W. and Rose, J. E., 1953, *Science* **118**:22.

Eccles, J. C. *et al.*, 1957, *J. Physiol.* (*London*) **138**:227.

Ferris, C. D., 1971, Electrode artifact in microelectrode systems, *Proc. 8th Ann. RMBS, Ft. Collins, Colorado*, pp. 27–32.

Ferris, C. D. and Stewart, L. R., 1974, Electrode-produced signal distortion in electrophysiological recording systems, *Trans. IEEE, BME* **21**:318–326..

Frank, K. and Becker, M. C., 1964, Microelectrodes for Recording and Stimulation, in *Physical Techniques in Biological Research*, Vol. 5 (W. L. Nastuk, ed.), Academic Press, New York.

Freygang, W. H., Jr. and Frank, K., 1959, Extracellular potentials from single spinal motoneurons, *J. Gen. Physiol.* **42**:749–760.

Gesteland, R. C., Howland, B., Lettvin, J. Y., and Pitts, W. H., 1959, Comments on microelectrodes, *Proc. IRE* **47**:1856.

Gray, J. A. B. and Svaetichin, G., 1951, Electrical properties of platinum-tipped microelectrodes in Ringer's solution, *Acta Physiol. Scand.* **46**:278–284.

Grunfest, H., Sengstaken, R. W., Oettinger, W. H., and Gurry, R. W., 1950, Stainless steel micro-needle electrodes made by electrolytic pointing, *Rev. Sci. Instru.* **21**(4):360–361.

Hittorf, J. W., *Pogg. Ann.* **89**:177 (1853); **98**:1 (1856); **103**:1 (1858); **106**:337 (1859).

Hubel, D. H., 1957, Tungsten microelectrode for recording from single units, *Science* **125**:549–550.

Ito, M., Kostyuk, P. G., and Oshima, T., 1962, Further study on anion permeability of inhibitory post-synaptic membrane of cat motoneurones, *J. Physiol.* (*London*) **164**:150–156.

MacInnes, D. A., 1961, *The Principles of Electrochemistry*, Reinhold, New York (Dover Publications Reprint).

Marmont, G., 1949, Studies on the axon membrane, *J. Cellular Comp. Physiol.* **34**:351–382.

Newman, J. S., 1973, *Electrochemical Systems*, Prentice-Hall, Englewood Cliffs.

Pratt, F. H., 1917, *Am. J. Physiol.* **43**:159.

Rush, S., Lepeschkin, E., and Brooks, H. O., 1968, Electrical and thermal properties of double-barreled ultra microelectrodes, *Trans. IEEE, BME* **15**:80–93.

Svaetichin, G., 1951, Low resistance microelectrode, *Acta Physiol. Scand.* **24**, *Suppl.* **86**:3–13.

Tasaki, I., Polley, E. H., and Orrego, F., 1954, Action potentials from individual elements in cat geniculate striate cortex, *J. Neurophysiol.* **17**:454–474.

Wolbarsht, M. L., MacNichol, E. F., Jr., and Wagner, H. G., 1960, Glass insulated platinum microelectrode, *Science* **132**:1309–1310.

CHAPTER 5

Half-Cells, Reversible and Reference Electrodes

When a metal is in contact with an electrolyte solution, a dc potential occurs which is the result of two processes. These are (1) the passage of metallic ions into solution from the metal, and (2) the recombination of metal ions in the solution with free electrons in the metal to form metal atoms. After a metal electrode is introduced into an electrolyte, equilibrium is eventually established and a constant electrode potential is established (for constant environmental conditions). At equilibrium, a dipole layer of charge (electrical double layer) exists at the metal–electrolyte interface. There is a surface layer of charge near the metal electrode and a layer of charge of opposite sign associated with the surrounding solution. Although diffuse, this dipole layer produces an effective electrical capacitance (C_p) which accounts for the low-frequency behavior of the electrode polarization impedance as discussed in Chapters 2, 3, and 4.

5.1. Half-Cell Potentials

When a metal is in contact with an electrolyte that contains its ions, the electrode potential developed is called a half-cell potential as the configuration with one electrode represents half of an electrolytic cell. Listed below are half-cell potentials for metals which can be used in physiological systems* :

$$Al = Al^{+++} + 3e^- \qquad +1.67 \text{ V}$$

$$Fe = Fe^{++} + 2e^- \qquad +0.441 \text{ V}$$

* Half-cell potentials for metal ion–metal interfaces at 25°C, from *Handbook of Chemistry and Physics.*

$$H_2 = 2H^+ + 2e^- \qquad 0.000 \text{ (Reference)}$$
$$Ag = Ag^+ + e^- \qquad -0.7996 \text{ V}$$
$$Pt = Pt^{++} + 2e^- \qquad -1.2 \text{ V}$$
$$Au = Au^+ + e^- \qquad -1.68 \text{ V}$$
$$Au = Au^{+++} + 3e^- \qquad -1.42 \text{ V}$$

In addition to metal–electrolyte interfaces, electrode potentials can be produced by ion transport through an ion-selective semipermeable membrane. In this case, the membrane is interposed between two liquid phases. Reversible transfer of a selected ion occurs through the membrane. For an ideal membrane, the developed electrode potential E is given by the Nernst equation

$$E = -\frac{RT}{ZF} \ln\left(\frac{C_1}{C_2}\right) \text{V}$$

where

F = Faraday constant, 96,495 C/mole

R = gas constant, 8.315×10^7 ergs/°K/mole

T = temperature, °K

Z = valence of ion involved

C_1, C_2 = concentration of ion on either side of membrane.

The Nernst equation predicts E accurately only in dilute solutions. For solutions normally encountered in physiological work, ion activities are used in place of ion concentrations. Ion activity, α, is defined by

$$\alpha = C\gamma$$

where C is ion concentration and γ is the activity coefficient for a given ion. In dilute solutions γ approaches unity and $\alpha \sim C$. The Nernst equation as modified by ion activities is thus

$$E = -\frac{RT}{ZF} \ln\left(\frac{\alpha_1}{\alpha_2}\right)$$

In practice, if we fabricate an ion-specific electrode by using a semipermeable membrane, the voltage (emf) E which we measure is dependent upon α_1 and α_2. If we wish to use the electrode to determine ion concentration C_1 or C_2, then we must know the appropriate activity coefficients γ. These are evaluated through the Debye–Hückel equations.

5.1.1. Debye–Hückel Theory

The derivation of the Debye–Hückel equations is not included here, but only the end results. A complete discussion is given by Bull (1964), Chapter 3. There are, however, several basic phenomena that we need to examine. There are three mechanisms which produce ions in water solutions. These are (1) solution of an ionic crystal, (2) oxidation of a metal or reduction of a nonmetal, and (3) ionization of a neutral molecule. Most metals when they ionize give up electrons to an electronegative element so that both acquire the electronic structure of a rare gas. Exceptions of interest in electrode work are iron, copper, silver, mercury, and zinc. In the ionized state, these metals do not acquire the completed outer shell structure of a rare gas and do have residual valences. They are then somewhat unstable and complex with various molecules more easily than do the stable ionized metals with completed shells. This accounts for the "poisoning" of silver–silver chloride electrodes and pO_2-measuring electrodes when used in high-protein environments such as blood.

When the alkali metals combine with chlorine they form ionic crystals; that is, the elements in the crystals exist as ions rather than as molecules. These crystals when dissolved in water form ionic solutions.

In aqueous solutions, neutral molecules tend to become ionized by two processes. Since H_2O has a large dipole moment bound to the ions of the molecule, the ions become hydrated and ionization is increased through the release of the energy of hydration. The high dielectric constant of H_2O (78) reduces interionic binding forces by $\frac{1}{78}$ of their values in air and promotes ionization of neutral molecules.

A useful description of an ionic solution (electrolyte) is ionic strength. The Lewis and Randall rule is stated, "In dilute solutions, the activity coefficient of a given strong electrolyte is the same in all solutions of the same ionic strength." For a heterogeneous solution, this rule is expressed mathematically as

$$S = \tfrac{1}{2} \sum_i C_i Z_i^2$$

where

$$C_i = \text{individual ionic concentration (moles/liter)}$$

$$Z_i = \text{valences of the respective ions}$$

As an example, the ionic strength of 0.1M NaCl is

$$S = \tfrac{1}{2}(0.1 \times 1^2 + 0.1 \times 1^2) = 0.1$$

The Lewis and Randall rule that strong electrolytes at the same ionic strength exhibit similar ionic effects is valid to $S \sim 0.1$. Solutions of higher strength exhibit anomalous effects. This is the basis for the work of Debye and Hückel.

The basic Debye–Hückel equation (at 25°C) is expressed as

$$\log_{10}\gamma = -\frac{0.509 Z_1 Z_2 S^{\frac{1}{2}}}{1 + 3.3 \times 10^7 a S^{\frac{1}{2}}} \tag{1}$$

where

$$a = \text{radius of the central ion}$$

$$Z_1 = \text{valence of cation}$$

$$Z_2 = \text{valence of anion}$$

$$S = \text{ionic strength}$$

$$\gamma = \text{activity coefficient}$$

The Debye–Hückel limiting law obtains for $S < 0.1$, which results in the simplified equation

$$\log_{10}\gamma = -0.509 Z_1 Z_2 S^{\frac{1}{2}} \tag{2}$$

In dilute solutions, the Debye–Hückel equations appear to yield exact results. In solutions of higher concentration ($S > 0.1$), there occurs a dip in the S–γ relation, as indicated in Figure 5.1. To account for the minimum in the S–γ plot, an additional term is added to the Debye–Hückel equation:

$$-\log_{10}\gamma = \frac{0.509 Z_1 Z_2 S^{\frac{1}{2}}}{1 + 3.3 \times 10^7 a S^{\frac{1}{2}}} - KS \tag{3}$$

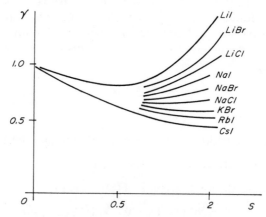

Figure 5.1. Mean activity coefficients γ of 1–1 halides at 25°C (redrawn from Bull, 1964, p. 76, with permission of author and publisher).

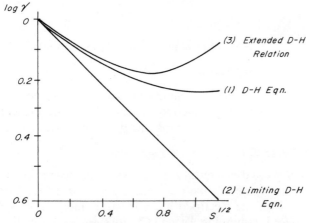

Figure 5.2. Plot of Debye–Hückel equations (redrawn from Bull, 1971, p. 77, with permission of author and publisher).

where K is determined experimentally and is called the *salting-out constant*. Approximate values for K and a are: $K = 0.1$; $a = 3.0 \times 10^{-8}$ cm. Thus,

$$-\log_{10}\gamma = \frac{0.509Z_1Z_2S^{\frac{1}{2}}}{1 + S^{\frac{1}{2}}} + 0.1S$$

Figure 5.2 presents the S–γ relations for equations (1), (2), and (3). The ramifications of Debye–Hückel theory and the relation

$$\alpha = C\gamma$$

will be discussed in Section 6.2.

5.2. Specific Half-Cells and Reference Electrodes

This class of electrodes is generally categorized as reversible or non-polarizable. By this, we mean simply that the electrode passes electric current without changing the chemical environment in the region of the electrode. The basic thermodynamic consideration for electrochemical reversibility is expressed by the Gibbs free-energy-loss relation

$$\Delta G = -nFE$$

for an electrolytic cell consisting of a metal–metal ion half-cell electrode and an ideal reference half-cell electrode. E is the cell emf and nF is the "capacity factor," where n is the number of faradays, F, transferred when the reaction proceeds to the amount $|\Delta G|$.

The basic relation which concerns us is

$$\text{oxidant}(Z) + (ne) \rightleftarrows \text{reductant}(Z - n)$$

where

$$Z = \text{valence}$$

$$n = \text{number of electrons transferred/mole}$$

$$e = \text{electron}$$

This relation expresses valence changes and accompanying electron transfers which occur in oxidation–reduction systems. The basic property of such systems is the conversion of chemical energy into electrical energy.

5.2.1. The Silver–Silver Chloride Electrode

A reversible electrochemical reaction occurs at the surface of the silver–silver chloride electrode. If the electrode is operated as one-half of a cell, Cl is deposited on the electrode when it is the anode, and AgCl on the electrode surface is reduced to Ag, and Cl^- ions are freed into the electrolyte solution when it is the cathode.

The Ag–AgCl electrode is generally represented (Janz and Ives, 1968) as

$$Ag|AgCl|Cl^-$$

It consists of a metallic silver substrate coated with AgCl, and is in contact with an electrolyte solution which contains a soluble chloride such as NaCl or KCl. Silver–silver chloride electrodes are reversible or nonpolarizing. This means that the electrode can pass electric current without changing the chemical environment in the vicinity of the electrode. Since AgCl is slightly soluble in H_2O, the electric currents at the electrode–electrolyte interface must be kept relatively small to maintain an unchanged chemical environment.

Under conditions of reversibility, the potential of a silver–silver chloride electrode is determined by two equilibrium relations:

$$Ag(\text{solid}) \rightleftarrows Ag^+(\text{solution}) + e^-(\text{metal phase})$$

$$Ag^+(\text{solution}) + Cl^-(\text{solution}) \rightleftarrows AgCl(\text{solid})$$

under the condition

$$\alpha_{Ag^+}\alpha_{Cl^-} = K_s$$

The ionic product must be equal to the solubility product. From the Nernst equation, we have

$$E = E^\circ_{Ag,Ag^+} + \frac{RT}{ZF}\ln\alpha_{Ag^+}$$

Using the relation

$$\alpha_{Ag^+}\alpha_{Cl^-} = K_s = 1.82 \times 10^{-10} \qquad (\text{Skoog and West, 1963})$$

then

$$E = E^{\circ}_{Ag,Ag^+} + \frac{RT}{ZF} \ln \left(\frac{K_s}{\alpha_{Cl^-}} \right)$$

$$= E^{\circ}_{Ag,AgCl} - \frac{RT}{ZF} \ln \alpha_{Cl^-}$$

where

$$E^{\circ}_{Ag,AgCl} = E^{\circ}_{Ag,Ag^+} + \frac{RT}{ZF} \ln K_s$$

and E°_{Ag,Ag^+} is the standard emf of the electrode for a given (reference) temperature $= -0.799$ V (Skoog and West, 1963).

5.2.2. Fabrication of Ag–AgCl Electrodes

Various techniques exist for producing Ag–AgCl electrodes, and a general survey is given by Ives and Janz (1961). A simple method for chloridizing a silver electrode is the following: An electrolytic cell is established as shown in Figure 5.3. The electrode, which should be clean high-purity silver, is made the anode. The electrolyte is not critical, and either dilute NaCl or HCl is frequently used. Nor is the cathode material critical; Offner (1967) even suggests copper wire though it is recommended that platinum wire or a platinum plate be used for the cathode. The concentration of the

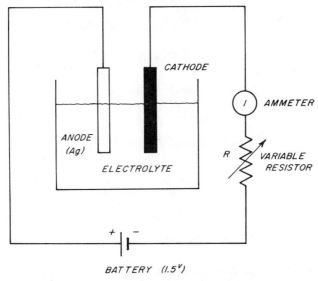

Figure 5.3. Apparatus for chloridizing a silver electrode.

electrolyte is not critical, although to insure a stable electrode, doubly distilled H_2O and reagent-grade chemicals should be used in making the electrolyte. The silver anode should be cleaned in *aqua regia* and washed with distilled H_2O. Recommended electrolytes are HCl in concentrations from 0.05N to 1.0N; KCl, 0.05N; and NaCl, 0.05N to 0.15N. Current in the cell (Figure 5.3) is limited to about 1 mA/cm^2 of anode surface by adjusting R. If $V = 1.5$ V dc, then R will be of the order of $1 \sim 10K \Omega$, depending upon anode size. The length of time that chloridizing is carried out determines the depth of the AgCl layer. Generally from 10 to 25% of the silver core should be converted to AgCl for stable electrodes. For small electrodes and wires, about 5 min will generally suffice. In any given situation, chloridizing time will have to be determined experimentally.

Since AgCl is photoreactive, the color of the finished electrode will depend upon the amount of light present during its preparation. Electrodes produced in the dark are usually dark-colored, either sepia or a plum shade. Silver–silver chloride electrodes produced in light are generally gray, tan, or pink. There are conflicting statements in the literature concerning color and the concomitant quality of the electrode.

If an aqueous solution of NaCl is used for the electrolyte, then the reactions which take place during the formation of the Ag–AgCl electrode are:

At the anode

$$Ag \rightarrow Ag^+ + e^-$$

$$Ag^+ + Cl^- \rightleftarrows AgCl$$

$$K_s = 1.82 \times 10^{-10} = \frac{[Ag^+][Cl^-]}{[AgCl]}$$

$$AgCl = solid \qquad (slightly soluble in H_2O)$$

$$\therefore [AgCl] = 1$$

Since $[Cl^-] \gg [Ag^+]$ essentially all of the Ag^+ is precipitated at the anode.

At the cathode

$$Na^+ + e^- \rightarrow Na^\circ$$

$$2Na^\circ + 2HOH \rightleftarrows 2NaOH + H_2 \uparrow$$

$$NaOH \rightleftarrows Na^+ + OH^-$$

Silver–silver chloride electrodes are sensitive to certain impurities in the electrolyte during their fabrication. Bromide is probably the most

TABLE 5.1. Response of Ag–AgCl Electrode to Interfering Halide Ions[a]

Indicated ion, M	Interfering ion, M	
	Br^-	I^-
10^{-2}, Cl^-	10^{-5}	—
10^{-1}, Cl^-	—	$< 10^{-8}$

[a]Data excerpted from Bishop and Dhaneshwar, 1963, p. 443.

serious contaminant, as shown in studies by Bishop and Dhaneshwar (1963) and reported by Janz and Ives (1968). As little as 0.1 % bromide in the chloridizing electrolyte is sufficient to reduce electrode life and cause failure. Table 1 summarizes this situation.

5.2.3. Standard Potential of Ag–AgCl Electrodes

The electrode potentials of half-cells such as the Ag–AgCl electrode are measured against the hydrogen electrode (which is discussed in Section 5.2.6.). The electrolytic cell configuration which is used in such determinations is shown in Figure 5.4. The standard potential of the Ag–AgCl electrode is determined from the emf of the hypothetical cell (IUPAC, 17th Conference, Stockholm, 1953) H_2 (ideal, 1 atm)|HCl (ideal, $m = 1$)|AgCl|Ag, with cell reactants in their standard states. After the measured values of cell emf have

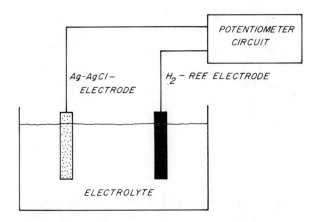

Figure 5.4. Apparatus for determining the standard potential of an electrode.

been determined from the experimental cell (Figure 5.4), the standard emf is calculated from

$$E = E^\circ - \frac{2RT}{F} \ln(m_\pm \gamma_\pm)_{HCl}$$

Rearranging terms, we obtain

$$E + \frac{2RT}{F} \ln m_\pm = E^\circ - \frac{2RT}{F} \ln \gamma_\pm$$

E° is determined by an extrapolation technique to zero concentration (see Ives and Janz, 1961, Chapter I, Section E). Bates and Bower (1954) give the value of E° as a function of temperature over the range 0°–95°C as

$$E^\circ = 0.23659 - 4.8564 \times 10^{-4}t - 3.4205 \times 10^{-6}t^2$$
$$+ 5.869 \times 10^{-9}t^3 \quad (t \text{ in } °C)$$

5.2.4. Aging of Ag–AgCl Electrodes

Even with very careful preparation of the electrodes, it is found that Ag–AgCl electrodes are initially unstable and that intraelectrode potentials occur. Stabilization can be accomplished by placing two electrodes in an

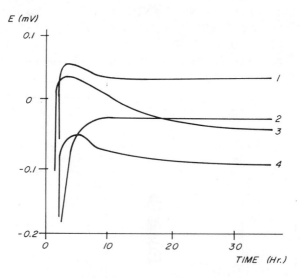

Figure 5.5. Electrode aging. Silver–silver chloride electrodes chloridized in 1N HCl and aged in 0.1N HCl. Electrodes 1 and 2 were coated twice (after Janz and Ives, 1968, p. 218).

electrolyte solution and connecting a short circuit between them. Electrodes normally stabilize in 24 to 48 hours, as indicated in Figure 5.5.

5.2.5. Use of Ag–AgCl Electrodes

There are several considerations in the use of Ag–AgCl electrodes. In making the electrical connection to the electrode one must be careful. If solder is used, it should be coated with a waterproof insulating material to prevent it from coming into contact with the electrolyte in which the electrode is immersed. This is also true of the connecting wire; otherwise, contamination of the electrode may occur because of chemical reactions between the solder or wire and the electrolyte.

Silver chloride is photosensitive (to ultraviolet light) and is so decomposed. It also produces a photovoltaic potential. Generally, Ag–AgCl electrodes should be stored in the dark and either used in subdued light or protected from light while in use. Excessive electrical noise produced by a given electrode may be indicative of light damage.

Silver–silver chloride electrodes require Cl^- ions for proper operation. When used in biological electrolytes, they have a sufficient supply of Cl^- ions available. If they are used as skin-surface electrodes in such applications as EEG or EKG recording, it is necessary to use a wetting solution or paste which contains Cl^- ions.

Silver–silver chloride electrodes are current-limited because they are reversible electrodes. Sustained passage of high direct currents results in either conversion of the electrode to pure silver (if used as a cathode) or conversion of all of the silver to AgCl (if used as an anode). Generally, these electrodes are used for signal recording, as opposed to stimulation, and operate into a high-input-impedance recording circuit. Electrode current is usually $< 10^{-9}$ A.

Because Ag–AgCl electrodes are thermodynamically reversible, they exhibit (after stabilization) low noise and theoretically zero electrode polarization impedance effects. They do produce a steady electrode potential, however, which produces a dc offset in direct-coupled systems. This requires compensation and is discussed in Chapter 7. This situation may cause problems in the sensing of low-level dc potentials. Because of the electrochemical nature of these electrodes, each one assumes an absolute potential. When two such electrodes are used as a sensing pair, a dc potential difference exists (frequently of the order of a few millivolts). If this potential difference remains constant, any measurement of a bioelectric potential is unaffected, except for a steady baseline elevation. In the usual case, however, the resting potentials of the two electrodes change unequally with time and environmental temperature. This results in objectionable baseline drift in experimental determinations.

Recently a new, and reportedly much more stable, solid silver–silver chloride pellet electrode has been introduced by Beckman Instruments, which does not appear to exhibit the undesirable qualities to the same extent as its fluid-bathed progenitors.

5.2.6. The Hydrogen Electrode

The hydrogen electrode is considered the primary standard to which other electrodes are compared. Unlike those of many other electrodes, the characteristics of hydrogen electrodes exhibit a high degree of reproducibility.

The basic electrode is an oxidation–reduction electrode operating under equilibrium conditions between electrons in a noble metal, hydrogen ions in solution, and dissolved molecular hydrogen. The activity of dissolved hydrogen, α_{H^+}, is taken as the independent variable and is fixed by maintain-

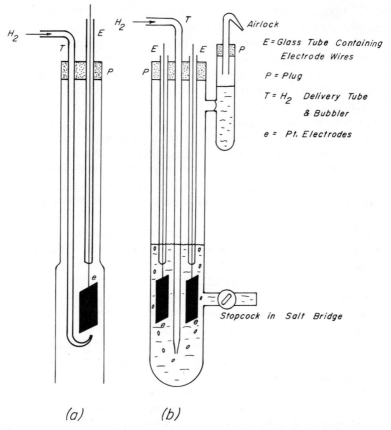

Figure 5.6. Hydrogen electrodes: (a) dipping electrode; (b) reference electrode.

ing equilibrium with a known partial pressure of hydrogen in the gas phase. Two typical electrodes are shown in Figure 5.6.

The electrode potential is given (Ives and Janz, 1961) by

$$E = E° + \frac{RT}{ZF} \ln\left[\frac{(\alpha_{H^+})^2}{P_{H_2}}\right]$$

By universal convention, $E°$ is set equal to zero, which establishes the hydrogen electrode as the standard reference. The expected exchange equilibrium for the hydrogen electrode

$$H_2(\text{aqueous solution}) \rightleftarrows 2H^+(\text{aqueous solution}) + 2e^-$$

is not established in the liquid phase. The metal associated with the electrode must catalyze the equilibrium. Thus the hydrogen electrode is represented by the following (Ives and Janz, 1961):

$$H_2(\text{aqueous solution}) \rightleftarrows 2H(\text{adsorbed on metal})$$

$$\rightleftarrows 2H^+(\text{aqueous solution}) + 2e^-$$

Various configurations have been developed for hydrogen electrodes and these are described in the literature (Ives and Janz, 1961). Platinum, gold, and palladium as well as other metals have been proposed for the metallic element of the electrode. Frequently platinized platinum is the preferred substance. The metal substrate may be wire, mesh, or foil. Preparation of the platinized electrode proceeds somewhat differently from the method outlined in Chapter 2. The platinum surface is usually cleaned in a solution of the following composition (Bates, 1954):

3 volumes	$12N$ HCl
1 volume	$16N$ HNO$_3$
4 volumes	H$_2$O

There are a number of technical considerations (Ives and Janz, 1961) concerning the quality of the hydrogen gas used, bubbling rate, barometric pressure, design of the bubbler (H$_2$ delivery tube), etc. The H$_2$ gas must be free of residual oxygen. This is generally accomplished by passing the gas through a deoxygenating furnace tube prior to its entry into the electrode. Gas is usually delivered to the electrode at a pressure which produces 2–3 bubbles a second. Newman (1973, Chapter 5) presents a detailed discussion of hydrogen electrode materials and preparation.

5.2.7. The Calomel Electrode

As in the hydrogen half-cell, there are a number of designs for the calomel half-cell, which uses mercury and calomel (mercurous chloride) as

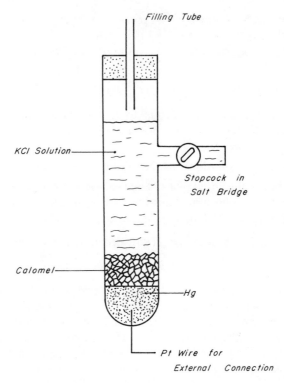

Figure 5.7. Construction of the calomel reference
electrode.

the reactants. In Section 5.4, the mercury–mercurous sulfate electrode,
used in standard emf cells, will be discussed. Calomel electrodes are used as
reference half-cells, especially in pH determinations.

The basic electrode is shown in Figure 5.7. Since calomel is relatively
insoluble, KCl solution is used as the electrolyte. The electrochemical
reactions which describe the half-cell are rather complex and are omitted
here. The emf of the calomel electrode depends upon the concentration of
the KCl solution. The following range of values can be expected:

Calomel half-cell emf at 25°C

KCl	*Emf (international volts)*
saturated	0.24
N KCl	0.28
0.1N KCl	0.33

As with all reference half-cells, there are special considerations in
fabrication to insure stability and reproducibility of the end product. The

mercury used must be specially purified and the KCl and calomel as free from impurities as possible. The calomel electrode is relatively easy to fabricate and, once constructed, is stable over long time periods. Special details in the design and preparation of these electrodes will be found in the literature; an excellent discussion with extensive bibliography is contained in Ives and Janz (1961), Chapter 3.

5.3. Salt Bridges

There are a number of instances in which electrical connection is necessary between parts of an electrode, between half-cells in a cell, between an electrode and a physiological system, or between an electrode and an electrolyte in which direct metallic connection is not permissible. In these cases, salt bridges are used. Let us examine several cases in which salt bridges have been used in electrodes previously described.

The electrolyte-filled micropipette is a typical example of a salt bridge. As noted in Chapter 4, there are cases in which it is not desirable to have direct metallic contact to a physiological system because of electrical considerations, toxicity, or trauma. In the micropipette a fluid coupling agent (salt bridge) provides the transition between the living system and the metallic electrode connection to associated electronics. Salt bridges are also used to reduce ac electrode polarization impedance effects (Schwan and Ferris, 1968). This application occurs incidentally in the glass micropipette. The pipette tip provides the small electrode contact area for compatibility with a physiological system, while the metal–electrolyte interface surface area can be made large (thus reducing current density at the interface) by using a relatively large metal electrode in the stem end of the pipette.

Salt bridges are useful for interconnecting two half-cells to form a reference cell, as shown in Figure 5.9, or for interconnecting a hydrogen reference half-cell with another half-cell for emf determination, as shown in Figure 5.8.

While salt bridges are usually fluid electrolytes such as aqueous solutions of KCl, NaCl, or $CdSO_4$, sometimes a gelatinous salt bridge is used. The gel should not be easily soluble in water at normal laboratory temperatures (20°–30°C). A recipe for a suitable gel salt bridge was given by Strong (1968): Soak 4 g of agar in 100 ml distilled H_2O for 8–12 hours. Heat the agar mixture in a beaker using a boiling water bath, not direct flame. Heat until agar has dissolved. Dissolve 30 g KCl in agar solution. If necessary, add just enough distilled H_2O to effect complete dissolution of the KCl. While the mixture is still hot, fill clean glass U-tubes. Care must be taken to fill the tubes entirely. Air bubbles must be excluded.

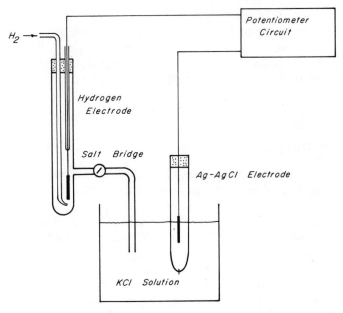

Figure 5.8. Use of a salt bridge in measuring the emf of an Ag–AgCl
electrode against a hydrogen reference electrode.

Agar salt bridges have been used for making contact to simple calomel
half-cells in *p*H determinations when platinum (quinhydrone) or antimony
electrodes are used. Glass and other membranes, as discussed in Section 6.3,
are now available for *p*H measurements, and gel salt bridges are generally
not used. The agar gel is subject to both chemical and bacterial contamination
and must be renewed periodically. Agar salt bridges must be stored in
contact with water or an electrolyte to prevent drying out. Impurities in the
agar may influence sensitive measurements.

Other types of salt bridges consist of electrolyte-saturated linen wicks
or bundles of camel's hair. These are useful when one wishes to minimize
the trauma to delicate tissues which occurs with electrode contact. Occasionally linen membranes are used in half-cells to maintain the physical position
of the reactants, especially in standard emf cell construction.

5.4. Reference Potential Cells—Standard Cells

In any potentiometric measurement (Section 5.5), it is necessary to have
a reference standard against which an unknown potential can be compared.
Several types of potential reference cells are used. Two common ones are
presented in Figure 5.9. One configuration consists of a mercury–mercurous

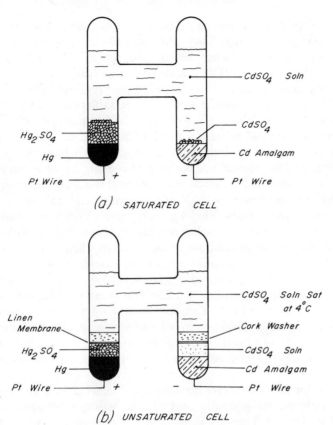

Figure 5.9. Standard emf reference cells.

sulfate half-cell connected by a cadmium sulfate salt bridge to a cadmium amalgam–cadmium sulfate half-cell. This type of reference standard cell is called the saturated (normal) cell as the $CdSO_4$ electrolyte contains crystals of the salt and is a saturated solution at room temperature (Figure 5.9a).

The emf produced as a function of temperature is given by the empirical relation

$$E_s(t) = E_{s20} - 0.0000406(t - 20) - 0.00000095(t - 20)^2$$
$$+ 0.00000001(t - 20)^3$$

where

$$E_{s20} = \text{emf generated at } 20°C$$

$$= 1.01830 \text{ international volts}$$

$$= 1.01864 \text{ absolute volts}$$

Normal cells in the configuration shown in Figure 5.9a are used as laboratory standards only. They are temperature sensitive and exhibit voltage hysteresis effects when heated and cooled. They are subject to mechanical shock and therefore cannot be transported easily.

The practical standard cell is the unsaturated type shown in Figure 5.9b. The $CdSO_4$ aqueous solution is adjusted to be saturated at 4°C and is thus unsaturated at usual laboratory temperatures. Unsaturated cells have the advantage that they can be transported and built in sizes small enough that they can be incorporated into electronic equipment which uses potentiometric circuits. They should not be handled roughly, however, and should be allowed to stabilize if subjected to mechanical shock or vibration.

Newly produced unsaturated cells generally yield an emf of 1.0190–1.0194 V abs (absolute volts). They age more rapidly than normal (saturated) cells, and the emf can be expected to decrease by about 30 μV/yr (0.003 %/yr).

There are certain precautions to be observed when using unsaturated standard cells. The permissible temperature range is $4°C < t < 40°C$. Standard cells should be protected against thermal shock, and all parts of the cell should be in thermal equilibrium. Currents in excess of 10 μA should not be drawn from standard cells. Current should be drawn only long enough to check galvanometer deflection in potentiometric circuits. Prolonged low-current drain, or short-term high-current drain, destroys standard cells. They are basically potential reference cells and are not capable of delivering power.

Except for the most exact measurements, zener diodes in a constant-temperature oven are generally used now as reference voltage sources in electrical equipment. They are stable and virtually immune to mechanical trauma.

5.5. Potentiometric Measurements

The potentiometer, used for accurate determinations of voltage in standards laboratories, has various applications in biological research. One, of course, is the accurate determination of various bioelectric potentials. In addition, many instruments can be constructed about the potentiometer as a central core. Figure 5.10 illustrates the basic potentiometer circuit. The instrument consists of two batteries, a standard reference cell and a working battery. In some instrumentation systems, the working battery is replaced by a regulated power supply, and a zener diode reference source is used in place of the standard cell. The heart of the instrument is a very accurately calibrated resistance, which is called a slide wire. It is usually in the form of a helix wrapped about a solid core. In operation, switch 1 is connected to the standard cell. The key (switch 2) is tapped at intervals

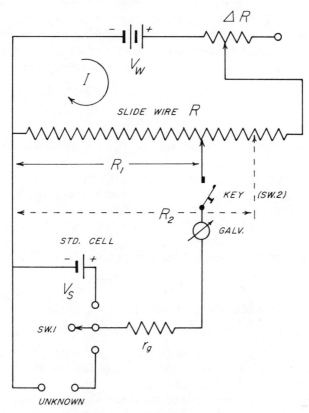

Figure 5.10. Potentiometer circuit.

while the sliding contact is adjusted. When the voltage drop across the section of the slide wire designated by resistance R_1 is equal to V_s, no deflection of the galvanometer is noted. The balance technique is a zero-current method and hence the resistance r_g of the galvanometer circuit need not be considered. The value of R_1 required for balance is recorded. The standard cell cannot be connected permanently, as under unbalance conditions, current exists in the galvanometer circuit. A steady current would damage the standard cell. This accounts for the insertion of the key.

Next, the galvanometer is connected to the unknown voltage and the slide wire again adjusted for balance with a resistance value R_2. The defining equations are

$$I = V_w/(R + \Delta R)$$

$$V_s = R_1 I$$

$$V_u = R_2 I$$

The first equation is not significant. The current I must remain constant and may be set at some convenient initial value by adjusting the rheostat ΔR. The working battery must be chosen so that the current drain upon it during operation does not reduce its terminal voltage. From the last two equations,

$$V_u/V_s = R_2/R_1$$

$$V_u = (R_2/R_1)V_s$$

Thus V_u is determined when the ratio (R_2/R_1) and the standard cell emf V_s are known.

A direct-reading instrument can be made by normalizing R_1 to $1\,\Omega$ and V_s to 1 V; then V_u (volts) $= R_2$ (ohms calibrated as volts). The slide wire is calibrated directly in voltage units. The rheostat ΔR can be used to zero-adjust the potentiometer for the required normalization. Many commercially available potentiometers are direct reading. Provision exists for adjusting initially to different standard cell emf's. A typical value for V_s is 1.019 V abs, when an unsaturated cell is used.

5.6. *p*H Electrodes and *p*H Meters

5.6.1. *The Antimony and Quinhydrone Electrodes*

A number of electrode configurations have been proposed for the measurement of pH. We will discuss several of them here. The electrodes which are the most easily fabricated are generally the poorest for pH measurements. These are the antimony and the quinhydrone electrodes. They are included here because of their simplicity of construction and are useful for self-study in pH determinations. The electrodes are shown schematically in Figure 5.11.

Both electrodes can be made from 10-mm-diameter soft glass tubing (although Pyrex is to be preferred for its stability). The tubing is drawn down to form a capillary, as shown in Figure 5.11. In making the antimony electrode, the end of the capillary is flame-sealed temporarily. The lumen is filled with small lumps of 99.8 % or purer antimony metal. The assembly is then heated sufficiently to melt the antimony. By gentle tapping of the glass and careful heating, it is possible to fill the tube with antimony without trapped air bubbles. To make the electrode connection, after the antimony has cooled, a piece of self-wetting (resin core) solder can be placed in the 10-mm end of the tube. This end of the tube is then heated until the solder melts and bonds with the antimony. While the solder is still hot, a piece of clean copper wire is introduced into the solder, which is then allowed to cool, thus forming the electrical connection. If desired, the wire may be tinned first with the same solder as used in the tube. The sealed end of the

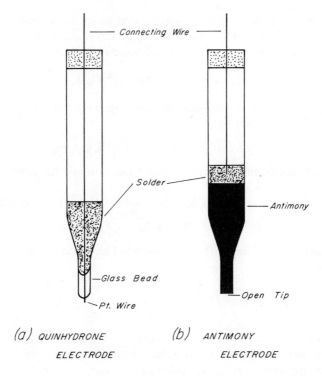

Figure 5.11. Simple *p*H electrodes.

tube is then broken off (cutting with an abrasive wheel is better) and the tip polished with crocus cloth. The electrode is now ready for use.

The quinhydrone electrode is of simpler construction than the antimony electrode. A piece of 24-gauge platinum wire about 1/2 in. long is introduced into the capillary, which is then sealed and beaded to the wire leaving a short tip exposed externally. The platinum wire should be cleaned in *aqua regia* briefly and washed in distilled water before using. After the bead has cooled, self-wetting solder is introduced into the lumen and the assembly is heated until the solder melts. After the solder has bonded to the platinum, but while it is still molten, a piece of clean copper wire is inserted into it to form the electrode connection when the solder cools.

Although these electrodes are easily constructed, they have a number of limitations. Antimony is toxic and tends to contaminate solutions being measured, especially when used in strong acids. The electrode tip tends to break down and must be reformed and polished.

The quinhydrone (platinum) electrode requires that a small amount of quinhydrone $[C_6H_4O_2 \cdot C_6H_4(OH)_2]$ be added to the solution under test.

This is necessary to insure hydronium ions in solution, since the electrode potential varies directly with hydronium ion concentration. In aqueous solution, quinhydrone exhibits the following redox behavior:

$$C_6H_4O_2 + 2H^+ + 2e^- \rightleftarrows C_6H_4(OH)_2$$

If the electrode is used against a hydrogen reference electrode, the cell so formed is described by

$$H_2(1 \text{ atm}), Pt|HCl(N) \vdots HCl(N), Q\cdot QH_2|Pt$$

(where the dotted line indicates a liquid junction and the solid lines metal–liquid junctions).

The electrode reaction is described by the following (Ives and Janz, 1961):

$$Q + 2H^+ + 2e^- \rightleftarrows QH_2$$

(reduction of quinone to hydroquinone). The electrode potential is thus given by

$$E = E^\circ + \frac{RT}{ZF} \ln\left(\frac{\alpha_Q}{\alpha_{QH_2}}\right) + \frac{RT}{F} \ln \alpha_{H^+}$$

Some of the advantages of the quinhydrone electrode are that it can be used in many solutions where other electrodes cannot be used, such as ethanol, methanol, acetone, and other organic and inorganic solutions, especially acids. It is not sensitive to barometric pressure. It is especially adapted to biological measurement (in systems which can tolerate the addition of quinhydrone) as only very small samples are necessary. The electrode can be made as a microelectrode if necessary (see Chapter 4).

At pH values ranging from 6 to 9, the electrode departs from ideal behavior and measurement errors occur. Quinhydrone electrodes tend not to be so long-lived as other electrodes. They are, however, mechanically rugged and can be used in extremely viscous fluids where glass electrodes (next section) would break. Electrical impedance is low.

5.6.2. The Glass Electrode

The electrode most commonly used in pH determinations is the glass membrane electrode shown in two configurations in Figure 5.12a,b. In operation, the glass electrode depends upon phase boundary potentials which form between the glass surface and the external electrolyte. Glass surfaces also acquire charge by ionic adsorption, although it is thought that this occurs to any degree only with excellent dielectrics such as quartz.

The considerable literature concerning glass electrodes has been concisely summarized by Ives and Janz (1961).

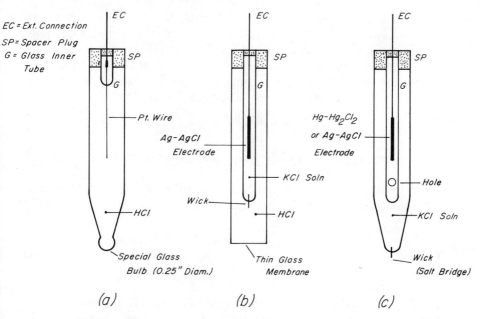

Figure 5.12. (a) and (b) Glass electrodes for pH determinations; (c) reference electrode.

The basic reaction which occurs at the *ideal* hydrogen reference electrode in a glass electrode–hydrogen electrode cell is

$$2H^+ + 2e^- = H_2$$

At 1 atm, the electrode potential is

$$E = \frac{2.3026RT}{F} \log_{10}\alpha_H$$

and

$$pH = -\log_{10}\alpha_H \quad (\text{since } \alpha_H < 1)$$

Hydrogen electrodes are difficult to use in many pH determinations and either Ag–AgCl or calomel reference electrodes are usually used. A typical pH-measurement reference electrode is shown in Figure 5.12c. Thus we establish a cell in which glass-electrode potential is compared with the potential of a nonhydrogen reference electrode which in turn has an electrode potential determined against a hydrogen reference. The cell which is formed by the two electrodes used in a pH measurement is then described by

$$H_2(g, Pt)|\text{solution } U \vdots KCl(\text{bridge}) \vdots \text{reference electrode}$$

The salt bridge is usually saturated or 3.5M KCl.

In current practice, the reference electrode is either calomel or silver–silver chloride and the hydrogen electrode is replaced by antimony, quinhydrone, or glass electrodes. The electrodes and pH-meter potentiometric circuit are calibrated against standard reference solutions. Under the assumption that liquid junction potential does not change if unknown solution U_1 is replaced by U_2, the Nernst relation would be

$$\Delta E = \frac{2.3026RT}{F} \Delta(-\log_{10}\alpha_H)$$

where $-\log_{10}\alpha_H = pH$ and ΔE = change in potential when U_1 is replaced by U_2.

If we now let one of the U's be a standard reference solution, we can substitute explicitly in the Nernst relation

$$E_u - E_{ref} = \frac{2.3026RT}{F}(-\log_{10}\alpha_{H,u} + \log_{10}\alpha_{H,ref})$$

$$pH_u = pH_{ref} + \frac{(E_u - E_{ref})F}{2.3026RT}$$

Glass electrodes are not quite linear, and the pH response of these electrodes depends strongly upon the type of glass which forms the bulb (Figure 5.12a) or the membrane (Figure 5.12b). Glass electrodes usually deviate from true pH values in the alkaline region. This is related to attack by strong alkalis upon the electrode and concomitant chemical breakdown of the glass. Alkaline pH errors appear to be ion specific and are related to sodium ions. In an attempt to combat this situation, lithium glass rather than soda-lime glass has been suggested (Ssokolof and Passynsky, 1932). Errors in pH reading also occur in the acid range for very low pH values. Various factors appear to be causes for reading deviations.

The glasses used for pH glass electrodes vary considerably in chemical formula. They have three general components in common, however. These are SiO_2, an alkali metal oxide, and the oxide of a bi- or trivalent metal. These are usually constituents in the values 60–75 mole %, 17–32 mole %, and 3–16 mole % respectively. The alkali metal oxide is frequently BaO, CaO, or SrO. Lanthanium oxide, La_2O_3, is frequently used as the bi- or trivalent oxide component. A typical pH glass might have the composition (mole %) SiO_2, 65 %; Li_2O, 25 %; Na_2O, 1 %; BaO, 7 %; La_2O_3, 2 %. This would be considered a lithia glass. Other oxides which have been used in pH glass formulas include TiO_2, GeO_2, and UO_2.

Different additives affect such factors as electrode resistance and alkaline- or acidic-reading linearity. It is difficult to find a single recipe which produces a satisfactory glass for readings of pH from 1 to 14. For best accuracy, at least two electrodes will be required to cover this range.

Glass is hygroscopic, and this generally leads to electrode breakdown in nine months to two years. Water tends to leach out alkali constituents in the glass, and the resulting lye breaks down and corrodes the membranes. On the other hand, the electrode must be hydrated in order to yield accurate pH readings. If glass electrodes are stored dry, they must be immersed in aqueous solution and allowed to stabilize before use. Generally, storage in distilled water is preferable and one must accept the fact that pH electrodes do not have long lives. Electrode life is directly related to membrane thickness, which ordinarily ranges from 50 to 140 μ. Deposits should not be permitted to form on the membrane as they are difficult to remove. Mechanical cleaning is impossible because of membrane fragility.

The reference electrode used in conjunction with a pH glass electrode is generally of the salt bridge type and a typical configuration is shown in Figure 5.12c. Use of a salt bridge (wick type in this case) protects the reference electrode from chemical action and contamination by the test solution. The electrode consists of an inner glass tube which contains mercury and calomel, which in turn is surrounded by an outer glass tube filled with saturated KCl solution. A small hole in the inner tube permits contact of the KCl solution with the mercury–calomel combination. Contact with the sample is usually established by either of two methods. In one case, a small fiber is sealed permanently into the immersion end of the tube. In the other, KCl reaches the sample by means of a small hole in the immersion end beneath a ground glass sleeve. A modification uses a ground glass plug in the immersion end. KCl solution seeps through the plug–tube juncture and makes contact with the sample. The calomel electrode is a practical reference electrode and is very stable with respect to potential, provided that temperature is reasonably controlled.

In some cases, the solution for which pH is to be determined cannot tolerate the calomel electrode. In such cases, a silver–silver chloride electrode is used. It is defined by

$$[Pt]H_2 ; HCl(N), AgCl; Ag$$

which in turn can be broken down as follows:

$$[Pt]; H_2, HCl \text{ (saturated with } H_2\text{)}; HCl \text{ (saturated with AgCl)}, AgCl \text{ (s)}; Ag$$

The electrode is constructed in the same manner as the calomel electrode, except that silver–silver chloride replaces mercury–calomel in the center tube.

There is a problem, however, with Ag–AgCl reference electrodes. As mentioned in Section 5.1, silver in the ionized state does not achieve the electronic structure of a rare gas. Because of this, when used in high-protein-containing solutions such as blood, Ag–AgCl electrodes become poisoned by adhesion of material to the electrode surface.

5.6.3. The pH Meter

A schematic diagram for a simple *p*H meter is shown in Figure 5.13. It is nothing more than a conventional potentiometer circuit. The precision slide wire (PSW) is calibrated in units of *p*H rather than in volts or ohms. A stable power supply is used, which may be a battery or a regulated electronic supply. The reference voltage source may be an unsaturated standard cell or a zener diode electronic reference.

In use, initial calibration of the meter is accomplished by setting the PSW to the calibration point (usually *p*H 7.0), the switch to CAL., and adjusting *R* for null indication of the meter. The switch is then set to USE and *p*H is determined by adjusting the PSW until the meter nulls. Should the PSW have to be recalibrated (different *p*H electrodes, etc.), buffered solutions of known *p*H are used.

The circuit shown in Figure 5.14 represents an electronic version of Figure 5.13. Electronic power supplies are used and an electronic voltage comparator with meter replaces the galvanometer.

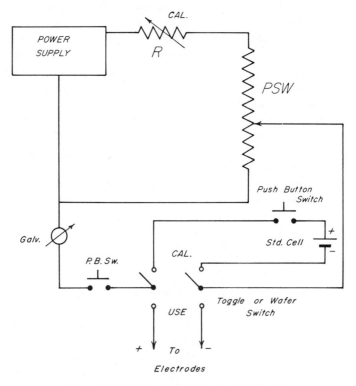

Figure 5.13. Circuit schematic for simple *p*H meter.

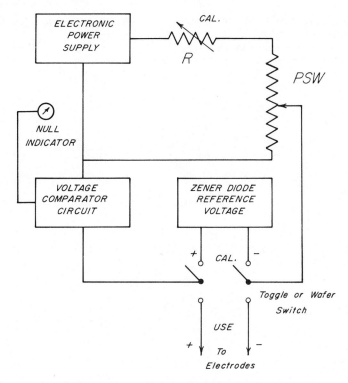

Figure 5.14. System schematic for electronic pH meter.

Many degrees of sophistication exist in modern pH meters. Some are direct reading and operate on the basis of a voltage offset against a standard voltage reference in the instrument.

5.7. References

Bates, R. G., 1954, *Electrometric pH Determinations*, John Wiley and Sons, New York.
Bates, R. G. and Bower, V. E., 1954, *J. Res. Nat. Bur. Std.* **53**:283.
Bishop, E. and Dhaneshwar, R. G., 1963, Silver- and halide-ion responsive electrodes, *Analyst* **88**:424–445.
Bull, H. B., 1971, *An Introduction to Physical Biochemistry*, 2nd ed., F. A. Davis, Philadelphia.
Hills, G. J. and Ives, D. J. G., 1951, *J. Chem. Soc.* 305.
Ives, D. J. G. and Janz, G. J., 1961, *Reference Electrodes*, Academic Press, New York.
Janz, G. J. and Ives, D. J. G., 1968, Silver, silver chloride electrodes, *Ann. N. Y. Acad. Sci.* **148**(1):210.
Newman, J. S., 1973, *Electrochemical Systems*, Prentice-Hall, Englewood Cliffs.
Offner, F. F., 1967, *Electronics for Biologists*, McGraw-Hill, New York.
Schwan, H. P. and Ferris, C. D., 1968, Four-electrode null techniques for impedance measurement with high resolution, *Rev. Sci. Instr.* **39**(4):481–485.

Skoog, D. A. and West, D. M., 1963, *Fundamentals of Analytical Chemistry*, Holt, Rinehart, and Winston, New York.
Strong, C. L., 1968, *Scientific American* **219**(3):232.
Ssokolof, S. I. and Passynsky, A. H., 1932, *Z. phys. Chem.* **A160**: 366.

CHAPTER 6

Ion-Specific Electrodes

In this chapter we examine the general problem of determining the amount of a specific ion in solution. Direct potentiometric measurement of ion activity is based upon the cell

Reference electrode	Salt bridge	Reference solution	Membrane	Test solution	Salt bridge	Reference electrode
E_1	E_2	E_3		E_4	E_5	E_6

where the E's represent boundary potentials. The simple theoretical basis for a membrane electrode is that energy-level differences exist between two different states of the same matter. The energy-level differences are proportional to the relative populations of the involved ions. For purposes of measurement with electrolytes, the energy-level differences are interpreted as electric potentials.

Most ion-specific electrodes operate on a potentiometric principle rather than an amperiometric or polarographic principle: that is, a change in potential, rather than a change in current, is sensed.

The number and variety of ion-specific electrodes is rapidly increasing with no end in sight. At the present writing, it is possible to use such electrodes to determine, either by direct or indirect measurement, ionic concentrations of the following species: ammonia, bromide, cadmium, calcium, chloride, cupric, cyanide, fluoride, fluoroborate, iodide, lead, nitrate, perchlorate, potassium, sulfide, sodium, sulfur dioxide, and thiocyanate, all by direct measurement, and by titration methods: aluminum, boron, chromium, cobalt, magnesium, mercury, nickel, phosphate, silver, sulfate, and zinc.

Historically the glass pH electrode was the first of the ion-specific membrane electrodes. In terms of the cell referred to above, the glass pH electrode is described by

Pt	HCl	Glass membrane	Test solution	KCl salt bridge	Ag–AgCl or Hg–Hg$_2$Cl$_2$ electrode

Although many membrane materials are available, glass is frequently used in biological work. Other membrane materials are discussed briefly in this chapter. Many sorts of membranes have been developed for testing and monitoring use in chemical manufacturing processes and are not directly applicable to physiological measurements. Many membranes function satisfactorily when only one ionic species is present. Biological electrolytes tend to be polyionic which limits, to some extent, the types of membranes which can be used. In single ionic systems, or where the differences in ion characteristics are pronounced, membrane electrodes can provide both a reliable and effective system for measuring ion concentration. Concentration is not measured directly, but rather is inferred from a measurement of ion activity.

In biological studies, we are normally interested in cationic determinations. The membrane–electrode potentials which are observed are described by the Nernst relation

$$E_{obs} = E° - \frac{RT}{nF} \ln(\alpha_{c^+})$$

where

E_{obs} = potential observed for a given activity of cation c^+, V

$E°$ = potential of ion being determined (when the energy of the cation is equal to the standard state, $E_{obs} = 0$), V

R = universal gas constant, J

T = temperature, °K

n = number of electrons transferred in the reversible reactions

F = Faraday's constant, 96,493 C/mole

α_{c^+} = activity of the cation under study

The membrane serves as a phase separator, which is made selectively permeable to those ions for which a concentration value (activity) is to be determined. Using the Nernst relation, with necessary correction factors for temperature and nonideality of the membrane, the potential between the known and unknown concentration can be measured and the unknown ionic concentration thus determined.

An accepted theory of ionic membranes is called the "fixed-charge theory" (Sollner, 1968). The pore walls of the membrane contain "inherently a definite number of potentially dissociable groups." Electronegative membranes contain anionic (acid) groups and electropositive membranes contain cationic (basic) groups. These ions are bonded in the membrane pore walls. The charge balance is maintained by "counter ions" of opposing charge. The nature of the "counter ions" is determined by the composition of the solution bathing the membrane.

If the solution ions are of one species, as in a concentration cell, the wall "counter ion" groups will be of the same type. If a polyionic solution is present, the concentration ratio of ions in the "counter ion" groups will depend on the individual concentrations in the half-cells and their respective adsorption properties. The adsorption properties which can be built into the membrane produce the selectivity.

It is the ability of membranes to pass certain ions selectively while retaining others that was first used by chemists. By doping the membrane with certain ions, the pores become lined with these ions. It is the interaction between these ions imbedded in the pore walls and those ions in solution, together with the extent of ion hydration, that determine what electrolytes are passed by a given membrane.

6.1. Special Considerations in the Use of Ion-Selective Electrodes

There are various problems which arise in the use of ion-selective electrodes. Specific electrodes measure activities rather than concentrations. Analytical methods for making the required conversion are available. For commercial electrodes, the manufacturer can supply the necessary information.

Some ionic species will interfere with other ionic species thus distorting the voltage reading produced by the electrode. Zinc interferes with the calcium electrode. Hydroxide interferes with the fluoride electrode. Sugar solutions produce low readings from pH and calcium electrodes that have been calibrated in aqueous standard solutions (Clarke, 1970).

To avoid problems of this nature, solutions should be prepared very carefully to avoid inclusion of interference ions, or buffered to adjust pH. Calibration should be carried out in solutions of the same general chemical composition as the solutions in which measurements are to be performed. Again, manufacturer's specification sheets should be consulted for information on particular electrodes.

Decomplexing agents can be added to some samples to render inert those ionic species which either interfere directly with electrode operation or complex with the ion being measured.

Ion-selective electrodes may be either of dipping or flow-through design. The former are used for general purpose, while the latter find specific application in anaerobic measurements.

6.1.1. Calibration Curves

Ion-specific electrodes should be calibrated periodically to insure that the electrodes are stable. This is usually accomplished by measuring the electrode potential in standardizing solutions produced by serial dilution. A calibration curve is then prepared using semilogarithmic paper. Electrode potential is plotted on the vertical linear scale while concentration is plotted on the horizontal logarithmic scale. If one desires to obtain concentration values for unknown solutions by direct measurement, then the ionic strengths of the standardizing solutions and the unknown samples must be similar.

Ionic strength adjustors can be added in equal amounts to both the standardizing and unknown solutions to damp out ionic strength differences. Generally a high level of a noninterfering electrolyte is added to produce a high but constant ionic strength in both classes of solutions.

Adjustments of pH may also be necessary for several reasons. Strong acidic or basic samples can damage sensitive membranes. In addition measurements of cyanide species must be done in basic solutions to prevent generation of toxic HCN gas.

Ion-specific electrodes are used in conjunction with reference electrodes. Frequently a pH reference electrode, such as a conventional calomel reference, should not be used, as KCl may not be a suitable electrode filling agent for many samples and in particular, for measurements of the silver ion. A reference electrode that is designed specifically for ion-selective measurements should be used.*

The techniques of "known addition" and "known subtraction" may be used with certain electrode instrumentation to avoid the use of ionic strength adjustors and calibration curves. These techniques are instrument-related and do not submit to general discussion. Orion (1973) treats this and other manufacturer-related electrode techniques.

A graphical method for determining concentrations has been developed which is also useful in titration determinations. This method, known as Gran's plots, involves the use of a proprietary plotting paper (Orion, 1973).

6.1.2. Use Extension of Existing Electrodes

In some cases ionic species for which no electrode exists can be measured using an electrode which is specific for another ionic species. Basically an indirect measurement is made which involves some chemical reaction of the ionic species of interest with the ionic species to which the electrode is selective.

Titration techniques may be applied so that a fluoride-selective electrode can be used to measure aluminum ionic strength.

---

*In some cases ammonium nitrate may be substituted as the filling electrolyte.

An iodide-selective electrode can be used to determine dissolved chlorine. Iodide is added when the sample solution is prepared. Loss of iodide (by reaction with chlorine) is then measured by the electrode.

6.2. Specific Membranes

The designation "membrane" is used here in a very general sense, as many materials are used. Commercially available electrodes include liquid membrane units, solid-state electrodes, glass membrane electrodes, and "plastic" membrane electrodes. General classes of ion specific electrodes in addition to those cited are: immobilized-liquid membrane electrodes, mixed-crystal membrane electrodes, enzyme electrodes, and antibiotic electrodes (Rechnitz, 1973). Certain of these membrane electrodes are now discussed in some detail. A generalized membrane electrode is illustrated in Figure 6.1a.

6.2.1. Clay Membranes

Clay membranes are prepared by the carefully regulated heating of layers of specially selected clays. By regulation of the doping levels, the clay matrix carries the necessary ions which regulate the adsorption of the ions, and hence selectivities. Clay membranes are limited in use, because they are difficult to prepare, have a high electrical resistance, and are easily broken.

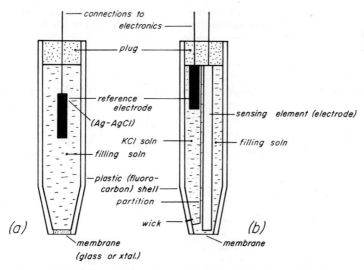

Figure 6.1. Membrane electrodes; (a) simple electrode, (b) general electrode (self-contained with built-in half-cell reference electrode).

Early clay membranes were thin plates formed from natural single crystals of chabazite and apophyllite (Marshall, 1939, 1942, 1944). These are zeolite (hydrous silicate) materials which possess a very open porous structure. Membranes of this sort are quite fragile and their fragility has led to the development of special clays (Marshall *et al.*, 1948).

6.2.2. Immobilized Liquid Membranes

The basic liquid membrane consists of some liquid ion-exchange resin which is restrained by an inert support. It suffers from several faults. The liquid–liquid interface is poorly defined and is subject to stirring effects and pressure differentials. It is mechanically fragile which can lead to mutual contamination of the two liquid phases.

The utility of liquid membranes as ion-selective electrodes lies in the mobility of their exchange sites. These are of molecular size, however, so that it is possible to immobilize an exchanger liquid in a bulk matrix. The constraint on the matrix is that it be permeable to microscopic charge carriers. Collodion is one such bulk material.

6.2.3. "Permselective" Collodion Matrix Membranes

Collodion is a mixture of pyroxylin in alcohol and ether. It represents an incomplete nitration of cellulose. Collodion matrix membranes can support high anionic or cationic charge densities and therefore can be made highly selective. Electrical resistance can be adjusted over several orders of

TABLE 6.1. Concentration Potentials (mV) across Permselective Membranes for 2:1 Concentration Ratios of KCl Solutions at 25°C [a]

Concentration ratio, 1-equivalent	E_{max}	Collodion membrane			
		Sulfonated polystyrene $\rho = 80\ \Omega\text{-cm}^2$	Oxidized $\rho = 300\ \Omega\text{-cm}^2$	PVMP[b] $\rho = 70\ \Omega\text{-cm}^2$	Protamine $\rho = 10.0\ \Omega\text{-cm}^2$
0.002/0.001	−17.45	17.25	17.20	−17.17	−17.17
0.004/0.002	−17.31	17.19	17.04	−17.15	—
0.01/0.005	−17.10	16.97	16.95	−17.09	−17.06
0.02/0.01	−16.86	16.74	16.74	−16.73	−16.76
0.04/0.02	−16.63	16.52	16.47	−16.48	−16.45
0.1/0.05	−16.30	16.10	15.80	−16.09	−16.06
0.2/0.1	−16.11	15.74	15.09	−15.73	−15.75
0.4/0.2	−15.95	15.40	13.90	−15.37	−15.03
1.0/0.5	−16.32	14.58	10.93	−14.48	−14.43

[a]Data extracted from Lewis and Sollner, 1959, p. 349; Gottlieb *et al.*, 1957, p. 157; Sollner and Neihof, 1951, p. 167; Neihof, 1954, p. 923.
[b]Poly-2-vinyl-*N*-methyl-pyridinium collodion membrane.

TABLE 6.2. Dependence of Concentration Potentials (mV) upon Concentration for 2:1 Ratios of KCl Solutions across Typical Permselective Protamine Collodion Matrix Membranes for Two Extrema of Membrane Resistance at 25°C [a]

Concentration ratio, 1-equivalent	Maximum theoretical potential	$\rho = 10 \ \Omega\text{-cm}^2$	$\rho = 2300 \ \Omega\text{-cm}^2$
0.002/0.001	−17.45	−17.17	−17.31
0.004/0.002	−17.31	—	−17.18
0.01/0.005	−17.10	−17.06	−17.10
0.02/0.01	−16.76	−16.76	−16.83
0.04/0.02	−16.63	−16.45	−16.55
0.1/0.05	−16.30	−16.06	−16.18
0.2/0.1	−16.11	−15.75	−15.91
0.4/0.2	−15.95	−15.03	−15.67
1/0.5	−16.32	−14.43	−14.86
2/1.0	−17.3	−12.33	−14.25

[a]Data extracted from Lewis and Sollner, 1959, p. 349.

magnitude. The collodion serves as a support matrix for the ion groups lining the pore walls and does not enter into the selectivity of the membrane. Pore size, however, is an important factor in selectivity and collodion pore size can be easily controlled. A disadvantage of collodion is its relative lack of resistance to alkaline solutions.

Tables 6.1 and 6.2 give examples of potentials developed across various collodion membranes. In Table 6.1, the membranes are doped with different ions. In Table 6.2, different resistances for one type of membrane are compared.

6.2.4. "Permselective" Noncollodion Membranes

Membranes using a matrix of materials other than collodion have also been prepared. "Heterogeneous membranes" are prepared by imbedding particles of ion-exchange resins in an inert binder, which serves as a frame to support the resin between the solutions.

"Homogeneous membranes" are also produced in which the ion-exchange resins are subjected to polymerization in a membrane form, that is, the resin is cased in a thin sheet. These sheets may be reinforced, for example, by casting over a woven fabric. A permselective membrane is shown schematically in Figure 6.2.

6.2.5. Liquid Ion-Exchange Membranes

Liquid ion-exchange membranes may also be used as membrane electrodes. These membranes function in a manner analogous to the

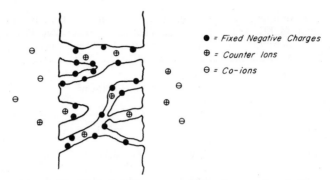

Figure 6.2. Diagrammatic representation for a permselective membrane (after Ives and Janz, 1961).

"permselective" membrane. Liquid ion-exchange membranes are produced by attaching the ion group whose concentration is to be measured to a suitable organic molecule of molecular weight from 300 to 600. An example is the secondary amine N-lauryl(trialkylmethyl)amine

$$CH_3(CH_2)_{10}CH_2N(H)C(RR''R')$$

where $R + R' + R'' = 11–14$ carbons. Other suitable compounds include tertiary amines, quaternary ammonium compounds, and various organic acid compounds.

These compounds are relatively insoluble in water and to form a membrane, are dissolved in water-insoluble solvents such as benzene, toluene, kerosene, or various higher alcohols. The water-insoluble fluid thus formed is used, in a thin layer, to separate the two aqueous solutions of differing ionic concentrations (reference solution and test solution). The organic solvent provides the barrier preventing mixing while the transport molecules carry the selected ions across the boundary.

6.2.6. Mixed Crystal Membranes

There are certain salt mixtures which are suitable as the membrane phase in an ion-selective electrode. Crystals such as LaF_3, Ag_2S, and the silver halides may be used directly as membranes, or they may be incorporated into an inert matrix. The active element in the sulfide-ion-selective electrode is polycrystalline Ag_2S. Transport of silver ions in the membrane forms the principle of operation (Hseu and Rechnitz, 1968).

A mixed-crystal membrane can be formed by adding CuS or PbS to Ag_2S, thus forming, respectively, $CuS–Ag_2S$ and $PbS–Ag_2S$ membranes. In the Ag_2S and $CuS–Ag_2S$ membranes, charge is transported by the movement of silver ions. Electrode potential is determined by the availability of S^{--}.

The $PbS–Ag_2S$ membrane has been applied in sulfate determination by titration with Pb^{++}.

The concept of a mixed-crystal membrane is not limited to sulfides and it should be possible to develop satisfactory two- and three-component crystal mixtures which can serve as suitable membrane materials (Rechnitz, 1973; Kummer and Milberg, 1969).

Mixed-crystal electrodes can function in both aqueous and nonaqueous media.

6.3. The Glass-Membrane Electrode

Practical ion-selective electrodes for analytical and clinical application generally use glass membranes. Glasses are used which are similar in composition to pH glass. This means that in use, glass electrodes for specific ion measurements have to be corrected for pH. The technique for doing this is presented below. The majority of the ion-selective electrodes which are available for biological work are cationic-selective and respond to sodium, potassium, calcium, lithium, or ammonium ions. Electrodes are also available which respond to the ions of cadmium, copper, lead and silver. Anionic-sensitive electrodes include those sensitive to Br^-, Cl^-, CN^-, F^-, BF_4^-, I^-, NO_3^-, and ClO_4^-.

The glass membrane electrode is described by the half-cell:

$$\begin{array}{c|c|c}
\text{Glass} & \text{Electrolyte} & \text{Ag–AgCl} \\
\text{membrane} & & \text{electrode}
\end{array} \rightarrow \begin{array}{c} \text{Connection to} \\ \text{potentiometer} \end{array}$$

The filling electrolyte varies in composition and is subsequently different for specific electrodes.

6.3.1. Calibration

As we have pointed out previously, electrodes respond to ion activity rather than ion concentration. In addition, there is lack of linearity between theoretical (from the Nernst relation) and measured values of emf for a given spread of ion activities for a particular ion in aqueous solution. Figure 6.3 presents this situation.

In the laboratory, a calibration curve is developed for each ion-selective electrode and the calibration is checked periodically to detect membrane degradation. From the Nernst relation, we can compute that the potential at the glass membrane changes by approximately 59 mV for each change by a factor of ten in ion concentration at 25°C.

To avoid the problem of activity versus concentration, a direct calibration of the electrode is carried out in known electrolytes. Typically one prepares known solutions by dissolving known amounts of a given salt in

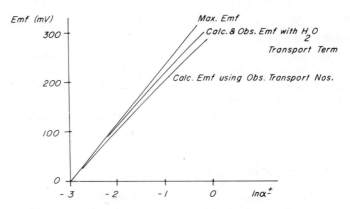

Figure 6.3. Experimental and theoretical emf values for membrane
cell (after Ives and Janz, 1961, Figure 2, p. 421).

doubly-distilled or deionized water. Commercial electrodes frequently
respond over the range 0.0001N to 1.0N. It is usually satisfactory to adjust
the concentration steps to 0.01N, 0.03N, 0.1N, etc. For each concentration,
the electrode emf is measured against a reference electrode (Ag–AgCl or

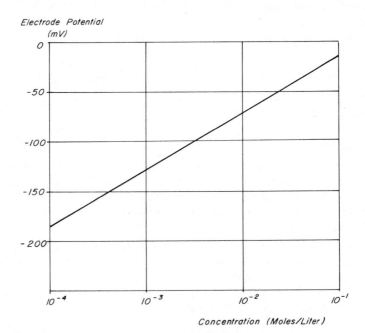

Figure 6.4. Typical working calibration curve for glass membrane sodium-
ion-selective electrode.

calomel) at constant temperature. The electrodes must be carefully rinsed with distilled water between samples. In this manner, an emf versus ion-normality calibration curve can be plotted as shown in Figure 6.4. For increased accuracy over a given range of concentrations, additional data points can be obtained by using additional reference solutions. In calibrating an ion-selective electrode and in using it later on, one must be sure that the entire active membrane surface is completely immersed in the test solution.

6.3.2. Correction for pH

Cationic-selective electrodes generally respond to hydrogen ions as well as the ion of interest. Thus one must adjust a test solution to suppress the influence of the hydrogen ions. This is achieved by adjusting the hydrogen-ion concentration so that it is at least four orders of magnitude lower than the lowest ion concentration to be measured.

As an example, let us suppose that sodium-ion concentration is to be measured over the range 0.0001M to 0.1M:

$$pNa = -\log_{10}Na^+$$

For 0.0001M sodium-ion concentration

$$pNa = -\log_{10}10^{-4} = 4$$

For 0.1M sodium-ion concentration

$$pNa = -\log_{10}10^{-1} = 1$$

Thus, if the lowest sodium ion concentration is $pNa = 4\,(10^{-4})$, the hydrogen-ion concentration must be suppressed to 10^{-8} or $pH = 8$.

This can be accomplished by adding a basic salt of different composition from the test salt. For example, if sodium is the test salt, calcium hydroxide could be added. One must be careful to use analytical grade reagents to avoid adding more of the test element to the sample. For convenience, one can simply saturate the test solution with the basic salt to assure pH suppression. Normally the test solution is modified for pH in the range 8–12 (see Figure 6.5). Some glass membranes are damaged by strong acids and alkalis as mentioned in Section 6.1.

Care of ion-selective glass-membrane electrodes is similar to that suggested in Chapter 5 for glass pH electrodes. One must be careful to maintain the correct amount of filling electrolyte.

Glass-membrane electrodes have a limited life, usually 1–2 yr if cared for. The glass becomes slowly weakened by contact with sample and standardizing solutions until it decomposes or shatters.

Some glass membranes which are selective for the potassium ion are also highly sensitive to acetyl choline. Thus electrodes made from this type

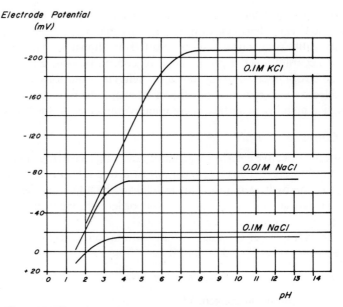

Figure 6.5. Effect of test-solution pH upon ion-selective electrode response;
sodium-selective electrode (typical response characteristic).

of glass should not be used in studies where acetyl choline is present (e.g., synapse studies in neurophysiological research).

Membrane electrodes, while they can be made to measure concentrations in the case of single ions, are rarely ion-specific in the more general polyionic situations.

Only when one ion species is highly hydrated, or otherwise large compared with other ions, or very insoluble in the membrane, will the membrane show a strong specific-ion selectivity. Selection between ions located near one another in the periodic table is seldom very good.

Development of ion-specific membranes of high selectivity has been slow, with both solid and liquid membranes being used. Other techniques, such as two-membrane chambers using differing membranes with somewhat different selectivity (not high selectivity but of significantly different ion permeability), require rather involved procedures and calculations. Response time of some of the membranes is also a limitation to their usefulness, especially with the advent of high-speed automatic wet chemistry analysis.

6.4. Examples of Some Ion-Selective Electrodes

Work by Lengyel and Blum (1934) and later by Eisenman *et al.* (1957) showed that addition of alumina (Al_2O_3) in quantities greater than 1 mole %

to glass formulations, resulted in a substantial increase in cation sensitivity of glass used as a membrane electrode. Experimentally it was noted that for a 10:1 increase in cationic concentration, a 58-mV increase occurred in electrode potential when the glass membrane electrode was measured against a saturated KCl calomel reference electrode at 20–22°C. The increase is 116 mV for an Ag–AgCl reference electrode. Alumina-glass electrodes *per se* are basically indifferent to the cation species present in a test solution.

6.4.1. The Sodium and Potassium Electrode

Studies by Eisenman *et al.* (1957) showed that by adjusting glass recipes, it was possible to find glasses that were ion-selective to a particular ion. They developed a sodium aluminosilicate glass of the following composition (mole %): Na_2O 11%, Al_2O_3 18%, SiO_2 71%. In a solution of pH 7.6, the differential sensitivity of such a glass membrane was measured at 250:1 for Na^+ over K^+. Analogous recipes can be developed to produce a potassium-selective glass.

Electrode structure is shown in Figure 6.6. It can be fabricated by bonding a short length of special sodium aluminosilicate glass to a piece of lead glass tubing. The end is then closed in a flame and blown to form the bulb. The filling electrolyte may be either 0.1N NaCl solution or 0.1N HCl. Electrical contact is made via a silver–silver chloride electrode (chloridized

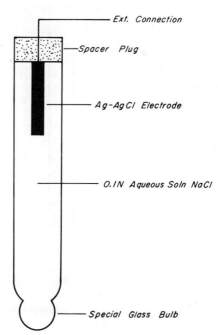

Figure 6.6. The sodium-ion-selective electrode.

silver wire). Normally a 3- to 10-mm-diameter bulb is used, although it is possible to make microelectrodes with bulb diameters less than $300\,\mu$. Lead glass is used because of its high electrical resistance. Bulb resistances have been measured at 10^8 to 10^{10} ohms. Alternatively, one can use the special glass for the entire structure and insulate all but the tip with a waterproof insulating material such as described in Chapter 4 for microelectrodes.

The cell for determining sodium-ion activity (calibrated as concentration) is described by

Ag–AgCl	0.1N KCl	Glass membrane	Test solution	Saturated-KCl calomel reference electrode

Ion-selective electrode

6.4.2. The Calcium Electrode

A calcium-ion-selective electrode was described by Ross in 1967. Its structural characteristics are shown in Figure 6.7. It uses a cellulose membrane (dialysis material) rather than a glass membrane. An internal liquid ion-exchange medium is used which is composed of 0.1M calcium salt of didecylphosphoric acid dissolved in di-*n*-octylphenyl phosphonate.

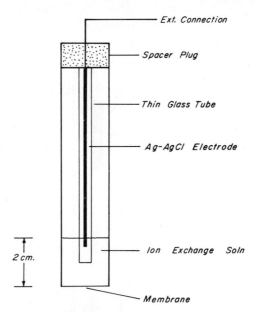

Figure 6.7. The calcium-ion-selective electrode. Ion exchange solution: 0.1M calcium salt of didecylphosphoric acid dissolved in di-*n*-octylphenyl phosphonate.

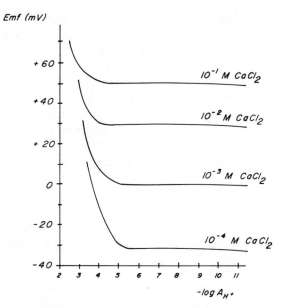

Figure 6.8. Effect of *p*H on calcium-ion-selective electrode for different calcium concentrations (redrawn with permission from Ross, 1967, Figure 1, p. 1379. Copyright 1967 by the American Association for the Advancement of Science).

Electrical connection is made via a salt bridge which is a small-diameter glass tube filled with 0.1M $CaCl_2$ aqueous 2% agar gel. A chloridized silver wire (Ag–AgCl electrode) is used for external connection.

The cell for Ca^{++} determinations is described by

| Ca^{++} electrode | Sample-solution liquid junction | Saturated-KCl reference electrode |

Experimental evidence (Ross, 1967) indicates that the calcium electrode can be used to measure free-calcium-ion activity in solutions which contain as high as 10^3 times as many sodium or potassium ions. In typical biological saline solutions, other monovalent cations exceed the uncomplexed-calcium-ion concentration by a factor of 10^2 at most. Figure 6.8 indicates electrode response as a function of *p*H.

6.4.3. Ammonia and Sulfur Dioxide Electrodes

Commercial electrodes of the general form shown in Figure 6.1b have been developed to measure ammonia and sulfur dioxide (Orion Research, Inc., Ann Arbor, Michigan, Models 95-10 and 95-64, respectively). These are self-contained units with an internal reference electrode. The ammonia electrode has a reported sensitivity of 20 ppb dissolved ammonia. The ammonium ion may be measured by adding a base to convert ammonium to ammonia. Nitrate is measured by adding a strong reducing agent to the sample. The electrode is reported to be virtually interference free, except for some volatile amines.

The ammonia electrode operates by the diffusion of dissolved ammonia from the sample through a highly gas-permeable membrane into an internal proprietary filling solution. Diffusion proceeds until a reversible equilibrium is established between the ammonia level of the sample and the level in the filling solution. Hydroxide ions are formed in the filling solution by the ammonia-water reaction

$$NH_3 + H_2O \rightleftarrows NH_4^+ + OH^-$$

The OH^- level in the filling solution which is detected by the sensing element, is directly proportional to the sample ammonia level. The electrode potential follows the approximate Nernst relation

$$E = E^\circ - 2.3 \frac{RT}{F} \log_{10}[NH_3]$$

where

$$E = \text{electrode potential, mV}$$

$$E^\circ \sim 180 \text{ mV}$$

$$2.3 \frac{RT}{F} = 59 \text{ mV/decade at } 29°C$$

The sulfur dioxide electrode works on exactly the same general principle as the ammonia electrode. Equilibrium is established between the sulfite level in the sample and the sulfite level in the proprietary filling solution. Hydrogen ions formed in the internal filling solution according to the reaction of SO_2 and H_2O

$$SO_2 + H_2O \rightleftarrows HSO_3^- + H^+$$

are sensed by the internal sensor. Electrode potential is given by

$$E = E^\circ + 2.3 \frac{RT}{F} \log_{10}[SO_2]$$

where the symbols have the same meaning as above, but E° may have a different value. Reported sensitivity is 0.10 to 1000 ppm SO_2 in aqueous samples only. The pH must be adjusted to 1.7, and interferences include hydrogen fluoride, acetic acid, and HCl greater than 1M.

6.5. The Oxygen Electrode

With the development of plastic membranes which are selectively permeable to the molecules of oxygen and carbon dioxide, but not to ions and water molecules, it has become possible to construct specific electrodes for pO_2 and pCO_2 measurements.

The basis for the operation of an oxygen electrode is that oxygen gas in solution reacts with a negatively charged (polarized) metal surface and forms OH^- radicals. Proteins, in particular, and other substances in solution, however, are also attracted to such surfaces and produce "poisoning" or reduced sensitivity to oxygen. Mattson and Smith (1973) have reported an experimental study of this phenomenon. In 1956, Clark developed a membrane electrode in which he used a polyethylene membrane to isolate the test solution from the metal portions of the electrode. The basic electrode configuration is shown in Figure 6.9. A platinum cathode is used and is formed from 10- to 25-μ-diameter platinum wire. Early electrodes used 2-mm-diameter wire, but it has been found that a smaller cathode requires less oxygen for operation and produces a smaller pO_2 gradient in the vicinity of the electrode (Severinghaus, 1968). This matter will be taken up subsequently. Normally the silver electrode is silver wire with a diameter usually larger than that of the platinum.

The silver wire is actually a silver–silver chloride reference electrode and is chloridized before the pO_2 electrode is used.

Electrode response time is relatively slow as one must wait for oxygen to diffuse through the membrane. The basic reaction at the cathode is probably

$$\tfrac{1}{2}O_2 + 2H^+ \text{ (in solution)} = HOH - 2e^-$$

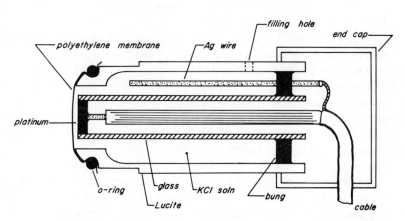

Figure 6.9. The basic Clark pO_2 electrode (see Clark, 1956, for original design).

TABLE 6.3. Gas Transmission Rates for Plastic Films 0.001 Inch Thick at 25°C (cc/24 hr/100 in²/atm)[a]

Film	O_2	CO_2	H_2O Vapor[b]
Polyvinylidene chloride	1.03	0.78	0.25
Monochlorotrifluoroethylene	1.5	16	0.04
Polyester (Mylar)	11.06	19.4	1.5
Cellulose acetate	110	560	90
Opaque high-density polyethylene	142	348	0.25
Polypropylene	187	—	0.7
Clear high-density polyethylene	226	1,030	—
Polystyrene	310	1,535	7.2
Low-density polyethylene	573	1,742	1.2
Tetrafluoroethylene	1,100	3,000	0.32
Ethyl cellulose	1,600	6,500	75
Silicone rubber (Dow S-2000)	98,000	519,000	170

[a]Data courtesy of Dr. R. John Morgan, Colorado State University, Ft. Collins, Colorado. Permeability data are generally directly proportional to thickness. For example, a 0.005 in. sheet would pass 1/5 to the above volumes. These values may change appreciably under different measurement conditions or as a function of manufacture.

[b]Water vapor transmission rate expressed as grams transmitted through 100 in.²/24 hr, with one side of film exposed to 90% relative humidity at 37°C and the other side at essentially 0% relative humidity by use of $CaCl_2$.

It is not unusual for the initial response time for an oxygen electrode to be as long as one minute. In general, the smaller the cathode tip, the faster the response. The membrane affects response time through its permeability to oxygen. Table 6.3 indicates values for different materials.

For blood-oxygen measurements, Eastman polypropylene has been found the most satisfactory material to use.

Oxygen electrodes do exhibit aging effects and these have several causes. The manifestations of aging are increased polarization current, increased response time, and sensitivity to hydrostatic pressure. Damage to the membrane or membrane aging is frequently a problem and is easily rectified by replacing the membrane. Another cause of aging is deposition of silver around the edges of the platinum cathode. This problem may be solved by carefully cleaning the electrode tip on wet Arkansas stone. Emery should not be used as it may imbed in the electrode surface.

The oxygen electrode is polarographic and in operation requires a polarizing potential of 0.2–0.9 V, with the platinum electrode negative relative to the Ag–AgCl wire. A circuit schematic is shown in Figure 6.10. Current through the cell is then a linear function of the oxygen tension in the solution which bathes the two electrodes. When the platinum is made

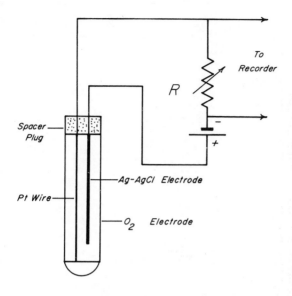

Figure 6.10. Simple polarograph circuit for pO_2 measurement.

slightly negative relative to the silver, oxygen reaching the platinum is reduced electrolytically. The reaction is not completely understood although a probable partial description has been given above. Normally about 0.2 V potential difference is sufficient. When the platinum is made more negative (-0.6 to -0.9 V) relative to the silver, the reaction rate of the electrolytic reduction is limited by the maximum rate at which oxygen can diffuse through the membrane to the electrode surface. At this point, the magnitude of the potential difference between the electrodes has little effect upon output current, and the output current variation is directly proportional to oxygen concentration in the bathing solution.

To decrease response time and prevent erroneous readings because of oxygen concentration gradients near the electrode, the test solution should be stirred, if possible. In this way, a homogeneous sample is assured. As is the case with other membrane electrodes, the oxygen electrode is pH sensitive. It is thought that at low pH levels, H_2O_2 may be formed rather than OH^-, thus using only half as many electrons per mole of oxygen. Oxygen-permeable membranes are also permeable to CO_2, and this gas freely diffusing into the solution bathing the platinum and silver–silver chloride electrodes alters the pH. The CO_2–pH effect can be eliminated by stabilizing the test solution at pH 7 using a pH 7 buffer stock solution to which 0.1M KCl has been added, or at pH 9 by using 0.5M $NaHCO_3$ and 0.1M KCl.

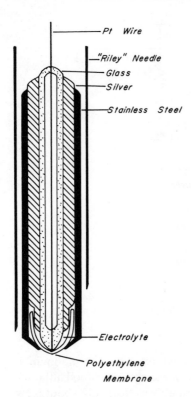

Figure 6.11. Microelectrode needle design for pO_2 measurements in tissues (the full electrode design is shown in Said *et al.*, 1961, p. 1130, Figure 1).

It is reported (Severinghaus, 1968) that the pH 9 buffer minimizes the CO_2 effect.

Various oxygen-electrode configurations have been developed. Figure 6.11 illustrates a miniaturized needle electrode for tissue and arterial use (Said *et al.*, 1961; Charlton *et al.*, 1963). Other workers (Fatt, 1968; Geddes and Baker, 1968) have stated that the platinum and silver wires should be exposed only in their cross-section. We have fabricated one such electrode as shown schematically in Figure 6.12a. The wire sizes are: Pt, $\frac{1}{64}$ in.; Ag, $\frac{3}{64}$ in. The platinum wire was enclosed in a snug-fitting drawn capillary tube and heated in a gas flame. Under microscopic examination the glass appeared to adhere tightly to the wire. The outer body of the electrode was made of a short glass tube of $\frac{5}{32}$ in. inside diameter. Epoxy glue was used to fill the tube. The face of the electrode was ground down with a stone to expose the two wires. Final polishing was accomplished by rubbing the electrode face against a wet glass microscope slide. The electrode was then placed in a small plastic box as shown in Figure 6.12b.

In this case (as is the case with the basic electrodes described by Fatt), no membrane is used. The chamber into which our electrode was placed

formed an oxygen-consumption system. The chamber was sealed against
entry of outside air. To determine a baseline for the electrode, the chamber,
except for an air space, was filled with distilled water. The chart record in
Figure 6.13 represents the electrode's response, which is somewhat erratic.
This points up several problems which are inherent in the design of oxygen
electrodes. We used a large cross-section platinum cathode. It was observed
that a bubble (presumably O_2) formed at the cathode when a potential was
applied to the electrode. Electrode current was dependent upon the rate at
which the bubble was reabsorbed into solution, thus permitting fluid contact
again with the cathode. If the test solution is saturated with O_2, the bubble
is not absorbed. The bubble, in turn, insulates the cathode from the test

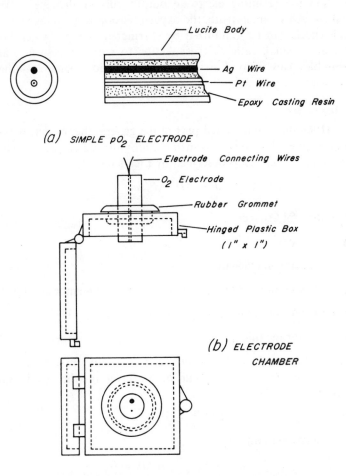

Figure 6.12. Simple pO_2 electrode and test cell assembly.

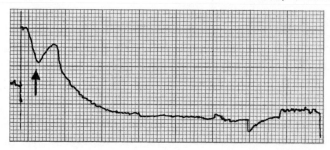

Figure 6.13. Performance of pO_2 electrode shown in Figure 6.11. Calibration: vertical 2×10^{-8} A/div.; horizontal 5 mm/min. Erratic behavior (arrow) is due to oxygen bubble.

solution and the resulting electrode output current indicates very little dissolved oxygen when, in truth, the oxygen tension is quite high.

Such effects can be minimized by stirring the sample, when possible, and by using a small cathode cross-section to reduce the surface area on which bubbles can form. Fatt (1968) recommends using 25-μ-diameter platinum wire.

6.5.1. Appendix

Fatt (1968) has summarized part of the general theory that pertains to oxygen electrodes. The oxygen transmissibility through a membrane which does not consume oxygen is described by Fick's law:

$$J = ADk\,\Delta P/\Delta x$$

where

J = flux, ml O_2/sec

A = area, cm^2

D = diffusion coefficient, cm^2/sec

k = O_2 solubility in the membrane, ml O_2/ml membrane/mm Hg

ΔP = O_2 tension differences across membrane, mm Hg

Δx = membrane thickness, cm

When oxygen bathes a layer of oxygen-consuming materials, such as a biological tissue, the oxygen profile in the layer (in the steady state) is described by

$$d^2P/dx^2 - Q/Dk = 0$$

with the boundary conditions

$$\text{at } x = 0, \qquad dP/dx = 0$$

$$x = L, \qquad P = P_a$$

where the additional symbols are

$$P = O_2 \text{ tension, mm Hg}$$

$$P_a = O_2 \text{ tension at open surface of layer, mm Hg}$$

$$Q = O_2 \text{ consumption rate, ml } O_2/\text{ml layer/sec}$$

$$x = \text{distance variable, cm}$$

$$L = \text{layer thickness, cm}$$

6.6. The CO₂ Electrode

If we combine the equilibrium equations for the first and second dissociations of carbonic acid, and invoke the principle of electroneutrality, the relationship between pH and pCO_2 can be developed (Severinghaus, 1968):

$$\alpha pCO_2 = [H_2CO_3]_P = \frac{A_H^2 + A_H A_{Na} - K_w}{K_1[1 + (2K_2/A_H)]} \tag{1}*$$

where

$$\alpha = \text{activity coefficient}$$

$$A_H = \text{hydrogen activity}$$

$$A_{Na} = \text{sodium activity}$$

$$K_1, K_2 = \text{dissociation constants}$$

$$K_w = \text{ion product of water}$$

This equation applies to an aqueous solution which contains $NaHCO_3$. In 1957, Stow *et al.* indicated the possibility of measuring pCO_2 by measuring the pH of a film of water separated from a test solution by a membrane permeable to CO_2. If the sensitivity of a membrane electrode based on this principle is defined by

$$S = \frac{\Delta pH}{\Delta \log_{10} pCO_2}$$

and Equation (1) is solved for different concentrations of $NaHCO_3$, it is found that $S = 0.5$ for distilled H_2O and $S = 1.0$ in aqueous solutions of 0.001–$0.1M$ $NaHCO_3$. CO_2 in pure H_2O produces equal numbers of H^+ and HCO_3^- ions. If HCO_3^- (by the addition of $NaHCO_3$) is already present in large quantity, then the change in HCO_3^- is negligible when CO_2 is introduced into the solution, but the change in H^+ is apparent.

Figure 6.14 illustrates how an electrode for CO_2 measurement might be constructed. It is essentially a glass pH electrode, bathed in an electrolyte,

*$[H_2CO_3]_P$ is the concentration of H_2CO_3 resulting from the partial pressure P of CO_2.

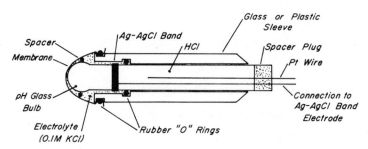

Figure 6.14. Electrode assembly for pCO_2 measurements.

and covered by a membrane freely permeable to CO_2. A silver–silver chloride reference electrode is incorporated into the system to form a full cell for direct pH measurement. In use, the electrode is dipped in the test solution and the output voltage read as pH on a pH meter. Since the electrode responds to the absolute value of pCO_2, it does not differentiate between gas or liquids of the same pCO_2 at constant temperature. Thus the electrode can be calibrated by flowing known concentrations of CO_2 in nitrogen or air over the electrode and calibrating the pH units on the pH meter in terms of pCO_2.

Various membrane materials can be used including rubber, silastic, Teflon, and polyethylene. Membrane thickness determines electrode response time, with thin membranes (0.001 in.) being the most responsive. Response time is similar to that of pO_2 electrodes (1–2 min) which is to be expected as the same sorts of processes are involved.

The electrical output from the pO_2 electrode is about 58–59 mV per 10:1 change in pCO_2 at 37°C (body temperature).

The most satisfactory filling electrolyte in terms of linearity of response and agreement between theoretical and experimental electrode behavior is $0.01M$ $NaHCO_3$ (or $KHCO_3$).

Measurement errors in the use of this electrode are related to membrane leaks (pin holes, etc.), inadequate temperature regulation during measurement, insufficient sample size, and air bubbles (in liquid samples) adhering to the outside of the membrane.

Combination electrode assemblies (pO_2–pCO_2) have been designed for clinical use in making blood-gas measurements (see, for example, Severinghaus, 1968).

6.7. Enzyme Electrodes

A main criterion for an ion-selective electrode is that it manifest charge transport and selectivity. Thus selective electrodes are not necessarily restricted to inorganic ions, or ions *per se*, provided the stated criterion can be met. Enzyme electrodes meet this criterion.

Enzymes may act as an intermediary agent in an electrode measurement system. Their actions are generally highly selective and the end products of their reactions are frequently simple ions for which selective electrodes already exist.

Guilbault and Montalvo (1969a, b) successfully produced an enzyme-electrode membrane by immobilizing an enzyme in a matrix which, in turn, was used to coat the active surface of a conventional cation-sensitive glass electrode. Their electrode was used to measure urea concentration. They fixed the enzyme urease in a layer of acrylamide gel. The gel was held in place against the glass electrode surface by nylon netting or a thin cellophane film. Urease acts specifically upon urea to produce ammonium ions. The ammonium ions, in turn, diffuse through the gel to produce a potential at the cation-sensitive electrode. This potential is directly proportional to urea concentration in the sample. This enzyme electrode was used for three weeks without loss of activity.

The chemical reaction in the urea electrode is

$$NH_2C(O)NH_2 + 2H_2O \xrightarrow{\text{urease}} NH_3 + NH_4^+ + HCO_3^-$$

It was reported that the electrode responded to urea concentrations from 5×10^{-5} to 1.6×10^{-1}M in tris(hydroxymethyl)aminomethane buffer. Electrode response was not independent of sodium and potassium ions when Na^+ was greater than one-half the urea concentration and when K^+ was greater than one-fifth of the urea concentration. Optimal electrode response time was 25 sec.

Guilbault and Hrabankova (1970, 1971) have described electrodes that are specific for L-amino acids and D-amino acids. Enzymes were immobilized by gel techniques (using acrylamide gel) on the tips of commercial cation-selective electrodes. L-Amino acid oxidase (L-AAO) was used for L-amino-acid-selective electrodes with the reaction

$$O^-C(O)CH(R)NH_3^+ + H_2O + O_2 \xrightarrow{\text{L-AAO}}$$
$$O^-C(O)C(R)O + NH_4^+ + H_2O_2$$

To prevent the release of carbon dioxide according to the reaction

$$O^-C(O)C(R)O + H_2O_2 \rightleftarrows OC(R)O^- + CO_2 + H_2O$$

catalase is added to the enzyme layer to catalyze the reaction

$$H_2O_2 \xrightarrow{\text{catalase}} \tfrac{1}{2}O_2 + H_2O$$

The total electrode reaction is

$$O^-C(O)CH(R)NH_3 + \tfrac{1}{2}O_2 \rightarrow O^-C(O)C(R)O + NH_4^+$$

D-Amino acid oxidase (D-AAO) has been immobilized in an acrylamide gel to produce D-amino-acid-selective electrodes. The reaction is

$$O^-C(O)CH(R)NH_3^+ + O_2 \xrightarrow{\text{D-AAO}} O^-C(O)R + NH_4^+ + CO_2$$

A recent survey article has summarized some additional enzyme electrodes and techniques (Gough and Andrade, 1973).

Aside from the criterion stated initially, there are several other factors which must be considered in the design and development of enzyme electrodes, or "biochemical-specific electrodes" as Gough and Andrade (1973) have defined them.

The electrodes should not evoke undesirable biological responses, such as tissue reactions, antigenic reactions, and the like. For use in physiological systems, they must be able to operate within normal life-system pH ranges. The enzymes used must be able to withstand entrapment within a synthetic hydrophilic gel, copolymerization with other enzymes, or physical entrapment between membranes.

Present enzyme electrodes, as is the case with most ion-selective electrodes, are potentiometric. This means that continuous measurements in the strict sense cannot be made. Polarographic or amperiometric systems, in which the measured ion species is continually removed or converted by the electrode, are the only true continuous-measurement systems.

Generally speaking, enzyme electrodes are highly specific, have reasonable response and life times, and hold considerable promise for further development.

6.8. Antibiotic Electrodes

Several antibiotics show distinct selectivity in their interactions with alkali metal cations (Pioda and Simon, 1969). Two antibiotics in particular, nonactin and valinomycin, associate with potassium ions in preference to sodium ions. It is difficult, by conventional methods, to fabricate an electrode which is selective to potassium in the presence of sodium ions. Using the preferential behavior of these two antibiotics, Simon (1969) was able to produce K^+-selective antibiotic membrane electrodes.

Nonactin was suspended in Nujol/2-octanol and valinomycin was suspended in diphenylether. These solutions were respectively incorporated into liquid membrane potentiometric electrodes. The usual commercial glass electrodes exhibit a K^+ to Na^+ selectivity on the order of 30:1. Simon's valinomycin electrode displayed a 3800:1 selectivity for K^+ to Na^+, and an 18,000:1 selectivity for K^+ with respect to H^+. Thus the electrode should be useful in strongly acid media in which glass electrodes cannot be used, or are ineffective.

The nonactin electrode exhibited a sensitivity of 150:1 for K^+ to Na^+ and a preferential selectivity for NH_4^+ over H^+ (Rechnitz, 1973).

6.9. Ion-Specific Microelectrodes

Walker (1973) has reported a technique for producing ion-specific microelectrodes which use liquid ion exchangers. The basic electrode is shown schematically in Figure 6.15. A liquid ion exchanger is formed by an organic electrolyte dissolved in a low-dielectric-constant organic solvent.

The microelectrodes which Walker reported were of two types: potassium sensitive and chloride sensitive. Tips of freshly pulled microelectrodes are normally hydrophilic. In order to coat the tip region with ion-exchanger material, the tips must be made hydrophobic. This is achieved by pulling borosilicate glass electrodes with tip diameters of 0.5–1 μ and immediately dipping the terminal 200 μ of tip into a fresh solution of 1% Clay–Adams Siliclad dissolved in 1-chloronaphthalene. After about 15 seconds, there will be a column of solution about 200 μ in length inside the tip. The electrode is then placed tip-up in a suitable holder and baked for 1 hour in an oven at 250°C. The electrode, after cooling, is ready for filling.

To make a K^+-sensitive microelectrode, the tip is dipped into Corning 477317 potassium exchanger, until there is a 200-μ column inside the tip. This requires from one to two minutes. Using a hypodermic syringe with a 30-gauge needle, 0.5M KCl solution is injected into the lumen. Details of this technique are given by Walker (1973). An Ag–AgCl sensing electrode is inserted into the assembly as shown in Figure 6.15, and the electrode is plugged or sealed with mineral oil.

A chloride-sensitive microelectrode is made in an identical manner, except that Corning 477315 chloride ion exchanger is used in place of the potassium ion exchanger.

These electrodes are suitable for intracellular determinations of K^+ and Cl^-.

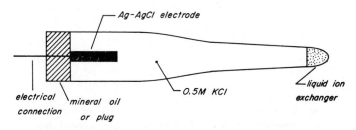

Figure 6.15. Ion-specific liquid ion exchanger microelectrode.

6.10. Conclusion

There is a wide variety of ion-specific electrodes currently available and it would appear that new techniques will develop rapidly in the future. A promising area is the synthesis of series of cyclic polyethers by Pedersen (1967). These selectively bind alkali metal ions (Izatt *et al.*, 1969). Such compounds hold promise for sensitive alkali-metal-ion-selective electrode membranes, and may well become the basis of some of the commercially available proprietary electrodes.

6.11. References

Charlton, G., Read, D., and Read, J., 1963, Continuous intrarterial pO_2 in normal man using a flexible microelectrode, *J. Appl. Physiol.* **18**:1247–1251.

Clark, L. C., Jr., 1956, Monitor and control of blood and tissue oxygen tensions, *Trans. Am. Soc. Artificial Internal Organs* **2**:41–48.

Clarke, M. A., 1970, The effect of solution structure on electrode processes in sugar solutions, *Proc. 1970 Tech. Session on Cane Sugar Refining Research*, p. 179.

Eisenman, G., Rudin, D. O., and Casby, J. V., 1957, Glass electrode for measuring sodium ion, *Science* **126**:831.

Fatt, I., 1968, The oxygen electrode: some special applications, *Ann. New York Acad. Sci.* **148**:81.

Geddes, L. A. and Baker, L. E., 1968, *Principles of Applied Biomedical Instrumentation*, John Wiley and Sons, New York.

Gottlieb, M. H., Neihof, R., and Sollner, K., 1957, Preparation and properties of strong base type collodion matrix membranes, *J. Phys. Chem.* **61**:154–159.

Gough, D. A. and Andrade, J. D., 1973, Enzyme electrodes, *Science* **180**(4084):380–384.

Guilbault, G. G. and Hrabankova, E., 1970, *Anal. Chem.* **43**:1779.

Guilbault, G. G. and Montalvo, J. G., 1969a, *J. Am. Chem. Soc.* **91**: 2164.

Guilbault, G. G. and Montalvo, J. G., 1969b, *Anal. Letters* **2**:283.

Hseu, T. M. and Rechnitz, G. A., 1968, *Anal. Chem.* **40**:1054, 1661.

Ives, D. J. G. and Janz, G. J., 1961, *Reference Electrodes*, Academic Press, New York.

Izatt, R. M., Rytting, J. H., Nelson, D. P., Haymore, B. L., and Christensen, J. J., 1969, *Science* **164**:443.

Kummer, J. and Milberg, M. E., 1969, *Chem. Eng. News* **47**(20):90.

Lengyel, B. and Blum, E., 1934, The behavior of the glass electrode in connection with its chemical composition, *Trans. Faraday Soc.* **30**:461.

Lewis, M. and Sollner, K., 1959, Preparation and properties of improved protamine collodion matrix membranes of extreme ionic selectivity, *J. Electrochem. Soc.* **106**:347–353.

Marshall, C. E., 1939, *J. Phys. Chem.* **43**:1155.

Marshall, C. E., 1942, *Soil Sci. Soc. Am. Proc.* **7**:182.

Marshall, C. E., 1944, *J. Phys. Chem.* **48**:67.

Marshall, C. E. and Ayers, A. D., 1948, *J. Am. Chem. Soc.* **70**:1297.

Marshall, C. E. and Eime, L. O., 1948, *J. Am. Chem. Soc.* **70**:1302.

Mattson, J. S. and Smith, C. A., 1973, Enhanced protein adsorption at the solid–solution interface: dependence on surface charge, *Science* **181**(4104):1055–1057.

Neihof, R., 1954, The preparation and properties of strong acid type collodion-base membranes, *J. Phys. Chem.* **58**:916–925.

Orion, 1973, *Orion Chemical Sensing Electrodes*, Orion Research, Inc., Ann Arbor, Michigan.

Pedersen, C. J., 1967, *J. Am. Chem. Soc.* **89**:7017.

Pioda, L. A. R. and Simon, W., 1969, *Chimia* **23**:72.

Rechnitz, G. A., 1973, New directions for ion-selective electrodes, in *Instrumentation in Analytical Chemistry*, pp. 224–229, American Chemical Society, Washington, D.C.

Ross, J. W., 1967, Calcium-selective electrode with liquid ion exchanger, *Science* **156**(3780): 1378–1379.

Said, S. I., Davis, R. K., and Crosier, J. L., 1961, Continuous recording in vivo of arterial blood pO_2 in dogs and man, *J. Appl. Physiol.* **16**:1129–1132.

Severinghaus, J. W., 1968, Measurements of blood gases: pO_2 and pCO_2, *Ann. N.Y. Acad. Sci.* **148**:115–132.

Simon, W., 1969, Paper presented at May, 1969, meeting of the Electrochemical Society, New York, N.Y.

Sollner, K. and Neihof, R., 1951, Strong-acid type of permselective membrane, *Arch. Biochem. and Biophys.* **33**:166–168.

Stow, R. W., Baer, R. F., and Randall, B. F., 1957, Rapid measurement of the tension of carbon dioxide in blood, *Arch. Phys. Med. Rehabil.* **38**:646.

Walker, J. L., Jr., 1973, Ion specific ion exchanger microelectrodes, in *Instrumentation in Analytical Chemistry*, pp. 232–239, American Chemical Society, Washington, D.C.

CHAPTER 7

Preamplifiers for Use with Bioelectrodes

The purpose of this chapter is to discuss certain important considerations in the design and use of preamplifiers, although a detailed presentation of circuit design techniques is beyond its scope. With microcircuits and chips becoming readily available at low cost, it will soon be unnecessary to design an amplifier system from scratch. There are, however, important considerations concerning input characteristics and dynamic response which are common to all amplifier systems. These are the topics which are presented here.

The signal input into a preamplifier may be a direct connection (dc amplifier) or the signal may pass through a capacitor which blocks a direct-current component in the signal (ac amplifier). Direct-coupled or dc amplifiers are more difficult to use since any direct-current component in the input signal tends to offset the operating point of the amplifier. Special design techniques are required to insure a stable system operation. Alternating-current amplifiers, on the other hand, do not suffer from dc-offset problems, but because of the input dc-blocking capacitor, their low-frequency response is limited and signal distortion can occur. A compromise is the chopper-stabilized amplifier which, in effect, converts dc signals to ac signals for amplification purposes and then converts the output back to dc.

A consideration in the use of amplifiers is the nature of the signal source with respect to a reference ground. Some signals may be processed on a single-ended basis (signal lead and ground lead as in recording from a single cell); other signals are processed on a difference basis (two signal leads and a reference ground as in electrocardiograph recording). In the first instance, a simple single-ended-input-circuit preamplifier suffices; in the second instance, a differential amplifier is required. These are some of the matters which are considered in this chapter.

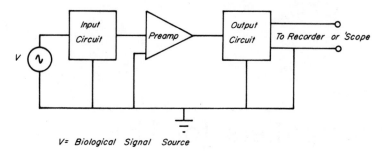

Figure 7.1. Basic preamplifier system.

The basic system under consideration is illustrated in Figure 7.1. The system shown is a simple single-ended preamplifier. Differential amplifiers will be considered as a separate topic.

7.1. Preamplifier Input Considerations

The basic input circuit for a general preamplifier is shown in Figure 7.2. A triangle is used to represent the amplifier and associated power supplies which provide the necessary bias voltages for its operation. In the case of a dc amplifier, C_c is replaced by a short circuit.

In the circuit, C_c represents the ac-coupling (dc-blocking) capacitor. C_i is the combination of shunt stray capacitance (from input cables, etc., to ground) and the dynamic input capacitance of the amplifier itself. R_i is the input resistance of the amplifier system and combines input bias resistance and dynamic input resistance of the active element in the amplifier.

The actual signal V_i which reaches the active circuitry of the preamplifier is modified from V_s by the input circuitry and is given by (Laplace transform

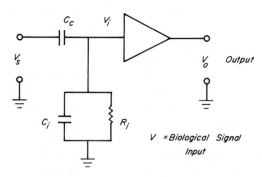

Figure 7.2. Generalized input circuit.

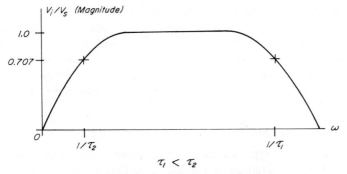

Figure 7.3. Frequency response of input circuit.

notation):

$$V_i = \left(\frac{sR_iC_c}{sR_i(C_i + C_c) + 1} \right) V_s$$

The input coupling circuit is effectively a band-pass filter or it can be considered as a differentiating circuit at high frequencies with the time constant $\tau_1 = R_iC_i$ and an integrating circuit at low frequencies with the time constant $\tau_2 = R_i(C_i + C_c)$.

The frequency response (V_i/V_s versus ω) of the input circuit is shown in Figure 7.3. From an examination of Figure 7.3, we see that we can expect signal distortion because of attenuation of both the high- and low-frequency components in the signal. At low frequencies, the signal components are shifted in phase which produces additional wave form distortion. The phase angle versus frequency for the input circuit is plotted in Figure 7.4.

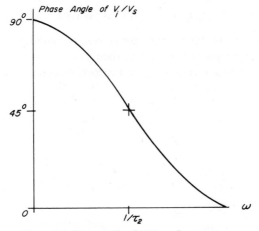

Figure 7.4. Phase angle of V_i/V_s vs. frequency ω.

When direct-coupled (dc) amplifiers are used, C_c is replaced by a short circuit. In this case, theoretically, there is no signal distortion. The R_iC_i combination, however, does represent an electrical load on the biological signal source and at high frequencies, when the shunting effect of C_i becomes important, there is attenuation of the high-frequency components. This is especially important in using micropipette electrodes, as described in Section 7.1.1.

7.1.1. Special Considerations for Microelectrodes

A rudimentary representation for an input circuit with glass micro-electrodes (using a dc amplifier) is shown in Figure 7.5. R_e represents the series resistance of the glass micropipette. A detailed description of electrode polarization problems was presented in Chapter 4. For this discussion it is sufficient to consider only the elements shown in Figure 7.5 since any signal distortion produced by electrode polarization can be incorporated into the description of V_s.

The voltage V_i presented to the preamplifier is

$$V_i = \left(\frac{R_i}{sR_iC_i + R_i + R_e}\right)V_s$$

The input circuit is a combined voltage divider (attenuator for all frequencies) plus a low-pass filter with time constant R_iC_i. If we neglect C_i for the moment, then

$$V_i/V_s = R_i/(R_i + R_e)$$

If we consider a typical commercial preamplifier such as is found in a high-gain unit in an oscilloscope, $R_i \sim 3 \times 10^6\ \Omega$. R_e is usually about $10^8\ \Omega$. Thus

$$V_i/V_s \sim 3 \times 10^6/10^8 = 0.03$$

With conventional amplifiers, the signal is attenuated by two orders of magnitude. Hence a signal of 100 mV at the tip of the micropipette would be only 3 mV at the input terminals of the preamplifier. Even when R_i is

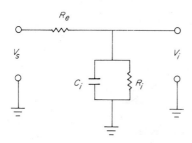

Figure 7.5. Simple microelectrode circuit model.

increased to $10^8 \, \Omega$,

$$V_i/V_s = 0.5$$

and half of the signal is lost. We see, then, that the input resistance R_i of a preamplifier to be used with micropipettes must be on the order of $10^9 \, \Omega$ or higher. This requires an electrometer input system which will be described subsequently.

If we now consider the effect of C_i we note another problem. Typical values of C_i are from 2 to 5 pF. Thus for $R_i = 3 \times 10^6 \, \Omega$, the minimum time constant is

$$\tau = R_i C_i = 3 \times 10^6 \times 2 \times 10^{-12} = 6 \times 10^{-6} \, \text{sec}$$

$$= 6 \, \mu\text{sec} \qquad R_e \sim 0, \text{ the metal microelectrode case}$$

For glass microelectrodes, R_e is large and the time constant is

$$\tau = R_e R_i C_i/(R_e + R_i)$$

$$= 5.82 \, \mu\text{sec} \qquad R_e \sim 10^8 \, \Omega$$

When R_i is increased to $10^9 \, \Omega$, the minimum time constant is 2 msec. This means that the 3-db cutoff frequency for the input circuit is only 80 Hz, so that if we increase R_i to prevent signal attenuation, C_i causes a substantial loss in frequency response and subsequent signal distortion. Figure 7.6 is a graphical illustration of this situation. The solution to this problem lies in using a feedback preamplifier system which reflects a negative capacitance at its input terminals, thus canceling the effect of C_i. This system is described in Section 7.7.1.

The discussion above indicates why glass micropipettes are generally used for low-frequency recording since they are characterized by the circuit shown in Figure 7.5. The low-pass filter aspect of Figure 7.5 accounts for this situation. Slowly varying signals are passed without appreciable distortion or attenuation (if R_i is sufficiently large). It should be noted that most of the

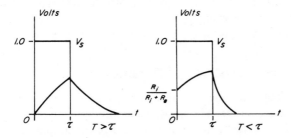

Figure 7.6. Pulse distortion by input capacitance.

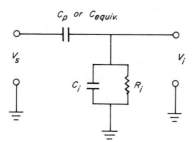

Figure 7.7. Basic equivalent circuit for metal microelectrode.

stray capacitances associated with micropipettes (Figure 4.10) can be lumped into C_i. Alternating-current electrode polarization in recording electrodes sometimes can be virtually eliminated by using a silver–silver chloride connection to the electrolyte in the lumen of the micropipette. When this is done, then Figure 7.5 is essentially a valid representation for a glass micropipette system.

When metal microelectrodes are used, Figure 7.7 indicates an approximate input circuit configuration. R_e is nearly zero and is omitted. C_p, the electrode polarization capacitance, now becomes important. When direct coupling to the amplifier is used, only C_p is present. For ac coupling, the series capacitance is given by

$$C_{equiv} = C_p C_c / (C_p + C_c)$$

Of course, C_p is frequency-dependent and its effect generally can be neglected above 1000 Hz. The circuit in Figure 7.7 is essentially that of Figure 7.2 and the mathematical description is similar except that C_p increases (on the series basis) as frequency increases and can be treated as a short circuit above 1000 Hz. A metal microelectrode system is basically a band-pass filter system, but with strong high-pass filter characteristics. When C_i is eliminated by using a negative-input-capacitance preamplifier, then it is a high-pass filter. This is why metal microelectrodes perform best for fast pulses such as found in neurophysiological processes.

We now see the reason for the general rule that glass microelectrodes are used for slowly varying processes (intercellular recording) and metal microelectrodes are used for rapid processes (neuron records).

7.2. Dynamic Response of Preamplifiers

In this section we consider such matters as frequency response, bandwidth, gain, and risetime. The basic system under examination is shown in Figure 7.8. We represent the active portion of the amplifier by a dynamic transfer factor g_m. The input circuit is defined by a transfer function $G_i(s)$

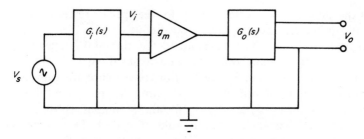

Figure 7.8. Abstraction of general preamplifier.

and the output circuit by a transfer function $G_o(s)$ such that

$$V_i = G_i(s)V_s$$

$$V_o = g_m G_o(s)V_i$$

where Laplace transform notation is used. The overall relation between the biological source voltage and the output voltage is then

$$V_o = G_i(s)g_m G_o(s)V_s$$

7.2.1. Amplifier Gain

If we designate V_i as the voltage available at the input terminals to the amplifier, then we define the gain of the amplifier through the relation

$$V_o/V_i = g_m G_o(s)$$

Since the factor $g_m G_o(s)$ can be complex if we substitute the steady-state variable $j\omega$ for s, we take the magnitude of the ratio V_o/V_i as the definition of gain. Thus

$$\text{Amplifier gain} = |V_o/V_i| = |g_m G_o(s)| = K$$

$$\therefore K = |g_m G_o(s)|$$

In the general case if V_o/V_i is a complex quantity, then $g_m G_o(s)$ can be represented as

$$g_m G_o(s) = A + jB$$

$$= \sqrt{A^2 + B^2} \, \angle\, \theta = \sqrt{A^2 + B^2} \, e^{j\theta}$$

where $\theta = \tan^{-1}(B/A)$.

The transfer characteristic of the amplifier (V_o/V_i) has now been represented in terms of a magnitude (gain) and a phase shift angle θ, such that

$$K = \sqrt{A^2 + B^2}$$

$$\theta = \tan^{-1}(B/A)$$

When the substitution $s = j\omega$ is made for the specific relations represented in A and B, we can make a gain-versus-frequency plot and a phase-angle-versus-frequency plot. These two plots define the total frequency response of the amplifier. Usually only the gain-versus-frequency plot is drawn. Typical plots are shown in Figure 7.9. For convenience, the gains have been normalized to unity. Normally over the amplifier pass band ($0 \leq \omega \leq \omega_c$ for the dc amplifier; $\omega_{cl} \leq \omega \leq \omega_{ch}$ for the ac amplifier), the phase shift angle θ is nearly constant and has the value 0° or 180°. Depending upon the number and arrangement of active elements in the amplifier, the output V_o is either in phase or 180° out of phase with the input V_i.

If we incorporate the input network $G_i(s)$ into the gain expression, then we can develop a relation for the ratio of the output voltage V_o to the biological signal source V_s.

$$V_o/V_s = g_m G_i(s) G_o(s) = \sqrt{A_s^2 + B_s^2} \angle \theta_s$$

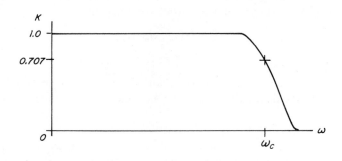

(a) DC AMPLIFIER

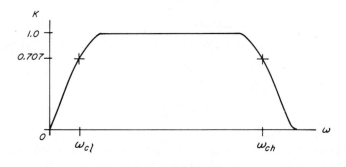

(b) AC AMPLIFIER

Figure 7.9. Normalized frequency response of an amplifier.

The overall system gain K_s is thus

$$K_s = |g_m G_i(s) G_o(s)|$$

K_s incorporates the effect of the input network as well as the effect of the amplifier and associated output network.

7.2.2. Bandwidth

A somewhat arbitrary definition is made for bandwidth. Amplifier cutoff is defined as that frequency (ω_c) at which the power output from the amplifier is one-half of the maximum power, or

$$|V_o/V_i|^2 = 0.5|V_o/V_i|^2_{max}$$

or

$$|V_o/V_i| = 0.707|V_o/V_i|_{max}$$

since $\sqrt{0.5} = 0.707$. Amplifier bandwidth is then described by the frequency range between the 0.707 or half-power points. A dc amplifier has only an upper half-power point, while an ac amplifier has two as shown in Figure 7.9. For the dc amplifier, the bandwidth (BW) is ω_c; for the ac amplifier BW = $\omega_{ch} - \omega_{cl}$.

Frequently the cutoff or half-power points are described in terms of decibel (db) units rather than voltage units. The conversion is

$$db = 20 \log_{10}(\text{voltage ratio})$$

or

$$20 \log_{10}(0.707) = -3 \, db$$

Thus the half-power points are called 3-db points, that is, signal level is 3 db down from the maximum level. The bandwidth defined by the half-power points is then the 3-db BW.

7.2.3. Gain-Bandwidth Product

If several amplifiers are connected in cascade, as shown for two stages in Figure 7.10, midband gain (amplification) is increased at the sacrifice of overall bandwidth. Note that the output circuit of stage 1 (g_{m_1}) has been incorporated with the input circuit of stage 2 (g_{m_2}). The overall system gain is given by

$$K_s = |V_o/V_s| = |g_{m_1} g_{m_2} G_i(s) G_o'(s) G_o(s)|$$

The gain-bandwidth product theorem states simply that in a single-stage amplifier, as gain is increased, bandwidth decreases such that the area under the frequency-response curve (Figure 7.9) remains constant. Simply: the

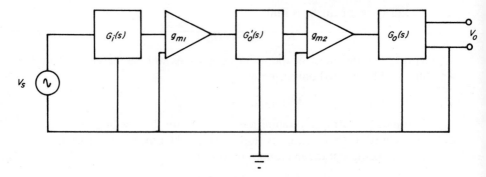

Figure 7.10. Two cascaded amplifier stages.

product of gain and bandwidth is a constant for a given amplifier configuration. Thus for a single amplifier stage as shown in Figure 7.8, if we increase the value of K, BW decreases such that

$$KBW = \text{constant}$$

In multiple stage amplifiers, as the number of stages is increased to increase gain, the bandwidth decreases to some extent generally, but not according to the gain-bandwidth-product relation.

7.2.4. Transient Response and Risetime

Experimentally, the transient response of an amplifier can be determined by applying an input voltage V_i of the form shown in Figure 7.11. This is called a step-voltage input and can be produced in the laboratory by a battery and switch.

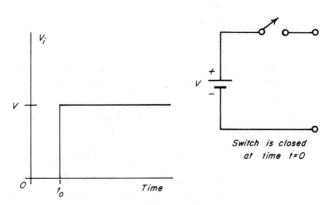

Figure 7.11. Step-function voltage.

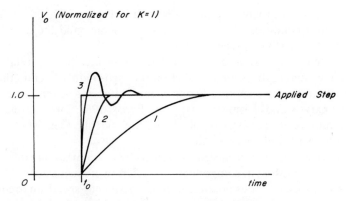

Figure 7.12. Transient response. Typical amplifier response to an applied voltage step. (1) Overdamped response, long response time to reach final value. (2) Critically damped response, minimum response time to final value without overshoot. (3) Underdamped response, minimum response time to final value, but oscillation ("ringing") occurs with overshooting of final value.

The output voltage V_o resulting from the application of the step voltage is displayed on an oscilloscope or chart recorder. Some typical response characteristics are shown in Figure 7.12. Normally the response voltage V_o will not be identical to V_i and will take one of the forms shown in the figure.

Amplifier risetime is defined in terms of response to a step-voltage input as indicated in Figure 7.13. There are various definitions, but usually it is taken as the time required for the output voltage V_o to rise from 10% of the final value to 90% of the final value for an applied step input V_i. With reference to Figure 7.13,

$$\text{Risetime} = t_2 - t_1$$

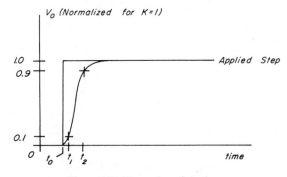

Figure 7.13. Illustration of risetime.

Referring to the transient response presented in Figure 7.12, we see that the underdamped condition (3) yields the fastest risetime and the overdamped condition (1), the slowest risetime.

Transient response and frequency response are directly related: the broader the bandwidth of an amplifier (at the high-frequency end), the more rapid is its transient response. Proof of this statement lies in the realm of Fourier analysis and is omitted here (see Ferris, 1962, for example). If we wish an amplifier to pass sharp pulses with high fidelity, then we must select an amplifier with wide bandwidth.

Any electrical signal can be represented by a summation (Fourier sum) of individual sinusoidal components of differing frequencies, amplitudes, and phase angles. It is thus possible to plot a frequency spectrum (frequency-response plot) for any voltage wave form. Generally, the frequency spectrum for a short pulse has the form shown in Figure 7.14.

If the frequency response of the amplifier is not equal to the BW of the frequency spectrum of the pulse, then wave form distortion will occur.

Generally, it is impractical to use step-function test signals, and repetitive square wave input test signals are used. Figure 7.15 is a diagnostic chart for some amplifier misdesigns relative to frequency response. Ferris (1962) gives a general mathematical treatment of these matters. A more practical treatment is given by Arguimbau and Adler (1956).

7.3. Alternating-Current, Direct-Current, and Chopper-Stabilized Preamplifiers

The input preamplifier which is connected to a pair of electrodes used for direct physiological recording has been called traditionally a "headstage." This terminology now appears on the wane. We have two choices of a basic amplifier system depending upon the type of signal that is to be processed.

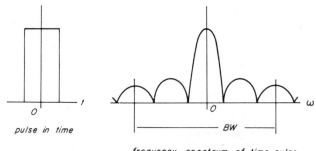

pulse in time

BW

frequency spectrum of time pulse

Figure 7.14. Frequency spectrum of a pulse.

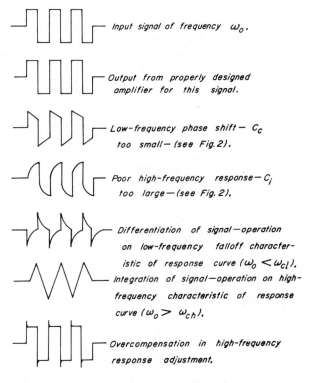

Input signal of frequency ω_0.

Output from properly designed amplifier for this signal.

Low-frequency phase shift — C_c too small — (see Fig. 2).

Poor high-frequency response — C_i too large — (see Fig. 2).

Differentiation of signal — operation on low-frequency falloff characteristic of response curve ($\omega_0 < \omega_{cl}$).

Integration of signal — operation on high-frequency characteristic of response curve ($\omega_0 > \omega_{ch}$).

Overcompensation in high-frequency response adjustment.

Figure 7.15. Square wave diagnosis.

Alternating voltages and fast pulses which possess little or no dc components are most easily amplified by an ac amplifier. Signals which are high in dc components and low-frequency components are processed by dc amplifiers.

There are several practical matters to be considered in selecting the amplifier to use. Theoretically a dc amplifier is suitable for all types of signals, but there are other factors which are important, and these are treated subsequently in this section.

It is not the purpose of this section or chapter to go into detailed amplifier design. The intent is simply to indicate when specific amplifier configurations are used and why. Circuits are presented only to illustrate specific points. Those interested in detailed design considerations should consult basic electronic engineering texts.

7.3.1. Alternating-Current Preamplifiers

Alternating-current amplifiers are recognized by the fact that input and output signals associated with the amplifier are coupled through capacitors

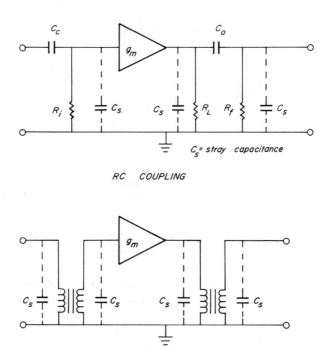

RC   COUPLING

TRANSFORMER   COUPLING

Figure 7.16. Alternating-current amplifier coupling techniques.

or transformers. Capacitive coupling is generally used as normally a greater operating bandwidth is possible than with transformers. Figure 7.16 illustrates two typical input-output (coupling) techniques. In this figure, the dotted-line capacitor connections represent stray capacitance. These are important only at very high frequencies or at very high impedance levels such as in glass-micropipette recording systems. The inherent problem with ac amplifiers is in the low-frequency response. Since, normally, capacitive (RC) coupling is used, this name being evident from the figure, we will consider that case alone. Two time constants are involved: $R_i C_c$ of the input and $R_f C_o$ at the output of the amplifier. The combination of R_i and C_c forms a simple high-pass filter and accounts for the low-frequency portion of the frequency-response plot shown in Figure 7.9b. The 3-db point at the low-frequency end is defined simply by

$$\omega_{cl} = 1/(R_i C_c) \qquad f_{cl} = 1/(2\pi R_i C_c)$$

if the time constant of the output circuit is much larger than the time constant of the input circuit. The combination also acts as a differentiating circuit for the falloff portion of the filter characteristic and produces the responses shown in Figure 7.15 for "C_c too small" and "$(\omega_o < \omega_{cl})$." The output coupling circuit with R_f and C_o also forms a high-pass filter and tends to increase the low-frequency falloff rate of the response curve.

The high-frequency falloff characteristic of an ac amplifier is produced by the interaction of the shunt capacitances (dotted lines in Figure 7.16) with R_i, R_f and R_L also in RC time-constant relations, but on the parallel rather than the series basis.

System risetime is affected by various factors. In micropipette recording systems, the RC combination of the electrode series high resistance (R_e in Figure 4.11) and shunt input capacitance C_i forms a low-pass filter, which on the falloff characteristic forms an integrator circuit. Wave shape is then altered as shown in Figure 7.15 for "poor high-frequency response" and "$(\omega_o > \omega_{ch})$." Since there is an inverse relationship between time and frequency, poor high-frequency response corresponds to poor risetime response. The risetime of an ac amplifier is also affected by the shunt capacitance in the output circuit. The capacitance generally arises from two sources: stray capacitance to ground as a result of the wiring layout, and internal capacitances associated with the active elements in the amplifier (tubes, transistors, or FET's). Risetime is equal to the product of the equivalent shunt resistance and the equivalent shunt capacitance in the output circuit, when there are no input circuit effects. Overall amplifier circuit gain K is directly proportional to equivalent shunt resistance in the output circuit. Thus we have a contradiction: to improve risetime, shunt R must be small; to maximize gain, shunt R must be large. Risetime can be related to the upper 3-db frequency cutoff point (ω_{ch}). In cases where the input circuit does not influence ω_{ch}, then risetime on a 63% basis (as opposed to the usual definition) $= 1/\omega_{ch} = R_\parallel C_\parallel$, where the values for R and C are the equivalent shunt values. If $R_\parallel$ is eliminated, the 63% risetime can be expressed in terms of g_m and the gain of the amplifier stage as

$$T_{63\%} = KC_\parallel/g_m.$$

The 63% figure arises from the response characteristics shown in Figure 7.12 for step-function response. In one time constant (RC), response attains 63% of its final value.

Alternating-current amplifiers generally do not perform well at low frequencies (< 10 Hz) and exhibit signal phase shift or signal differentiation. High-frequency response must be selected so that risetime is not compromised for the signals being processed. On the other hand, ac amplifiers provide stable operation, are relatively free from baseline drift, and require much less calibration than dc amplifiers.

7.3.2. Direct-Current Preamplifiers

Direct-current (direct-coupled) amplifiers are required when dc signals, slowly varying ac signals (< 10 Hz), and certain low-repetition-rate (e.g., EKG) signals are processed. In an ac amplifier, the capacitors C_c and C_o isolate the internal dc bias voltages, necessary for proper amplifier operation, from external dc signals. With dc amplifiers, the external dc signals are those which are being processed, and since these add or subtract from the bias voltages necessary to amplifier operation, dc amplifiers then have shifting operating points. This situation causes instability, baseline drift, and calibration difficulties. Although many instruments use dc amplifiers, their general use should be avoided in favor of chopper-stabilized amplifiers. The cost factor is the reason that conventional dc amplifiers are in common use.

7.3.3. Direct-Current Offset

In order to avoid shifting the operating point of a dc amplifier beyond its linear range, it is frequently necessary to apply a dc bucking voltage at the amplifier input to offset (neutralize) the input signal. For example, suppose that we use a silver–silver chloride recording electrode and a platinum indifferent (reference) electrode in a system such as is shown in Figure 4.8. In the quiescent situation at constant temperature, we can expect a small, steady dc potential at the amplifier input terminals. A given dc amplifier might be designed to process signals which vary ($+$ or $-$) about a 0-volt baseline. The electrode offset potential, in this case, would present a nonzero baseline. To correct this offset to zero, a simple circuit as shown in Figure 7.17 might be used. In commercial instruments, a more complicated system than that shown would probably be used. The "dc" offset adjustment

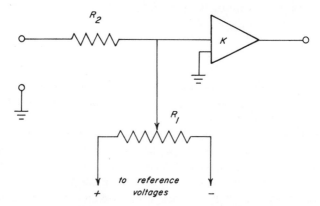

Figure 7.17. Simple dc offset circuit.

on many commercial dc amplifiers and recorders simply provides baseline restoration for offset potentials produced by electrode systems and other sources. The use of such an offset control to restore an operating baseline does not affect amplification of signal variation about the baseline, but does introduce a varying attenuation of the input signal depending upon the value of R_1. The R_1–R_2 combination forms a voltage divider.

Another problem associated with dc amplifiers (and ac amplifiers as well) is signal saturation. All amplifiers possess a certain dynamic range, which may or may not be symmetric. Alternating-current amplifiers generally have a symmetric dynamic range. Direct-current amplifiers may be designed to accept a symmetric signal ($\pm V$ volts) or they may accept only positive signals (0 to $+V$ volts) or only negative signals (0 to $-V$ volts). Some dc amplifiers may be asymmetric in the form that they accept signals of the form $-V_1 \leq V_{in} \leq +V_2, |V_1| \neq |V_2|$. When the input voltage limits are exceeded, saturation generally occurs; that is, the output voltage no longer increases linearly as input is increased. In hard saturation, the output remains constant as input is increased.

Generally, ac signals are symmetric about a zero-volt baseline. If a dc amplifier is used to process such a signal, it may be necessary to use a dc offset to put the ac signal into the operating range of the dc amplifier. This is also true of an ac signal which possesses a dc component.

7.3.4. Chopper-Stabilized Preamplifiers

At present, the best compromise between the stability of the ac amplifier and the low-pass properties of the dc amplifier is the chopper-stabilized preamplifier. The input-dc or low-frequency signal is fed through a chopper, which is simply an electrically actuated switch which samples the input signal at a predetermined fixed rate. The input signal is converted into a train of pulses whose amplitudes are proportional to the voltage level of the original signal. The output of the chopper is then an alternating voltage which can be amplified by a conventional ac amplifier. The ac amplifier usually is designed to have a narrow bandwidth centered about the frequency of operation of the chopper. In this manner, wide-band noise is reduced when the signal is amplified. The chopping action does inject some switching (commutator) noise into the signal.

Both mechanical and electronic choppers are used. The mechanical units frequently switch at the power-line frequency (60 Hz) and through the arrangement of the switch contacts produce a chopped signal at the second harmonic (120 Hz). Mechanical choppers consist usually of a set of fixed-switch contacts, an armature (moving reed with contacts), and an electro-magnet assembly which actuates the armature. An operating frequency of 60 Hz, although convenient, is not desirable because of possible hum pickup in the signal. Higher-frequency mechanical choppers can be designed to operate from electronic oscillators. There is, however, an upper limit on the rate at which mechanical switching can take place reliably without "contact bounce" and other problems which cause signal distortion or loss of signal.

Electronic choppers incorporate semiconductor components in circuit configurations which are analogues of mechanical choppers. Chopper frequencies are usually of the order of 1–10 kHz and narrow bandwidth (tuned) ac amplifiers are used. Photoelectronic choppers are also used. Figure 7.18 illustrates two types of chopper input circuits.

Chopper amplifiers do have several disadvantages. Mechanical choppers have a short lifetime and are expensive. In general, this system is satisfactory for dc signals and slowly varying ac signals. Unless quite sophisticated circuitry is used, chopper amplifiers are not satisfactory for processing repetitive

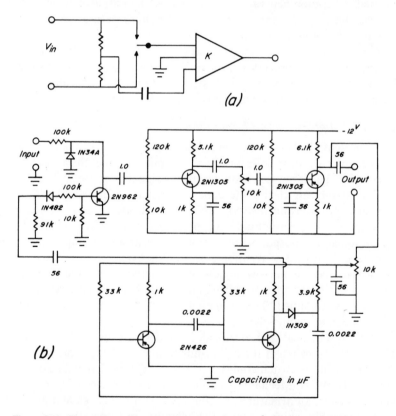

Figure 7.18. Chopper amplifier circuits; (a) mechanical chopper input, (b) electronic chopper, multivibrator driver, and amplifier (Ferris, 1965*b*).

pulsed signals such as EKG signals. Unless the chopping rate is very high, portions of the leading and trailing edges of such pulses, and frequently the peak value, will be lost.

After amplification, the chopped signal can be converted back to dc by using a demodulator circuit which is simply a rectifier and low-pass filter. Mechanical demodulators are also used and consist of additional switch contacts actuated by the same armature which is used to chop the input signal. In this manner, synchronous operation is assured.

Advantages of chopper amplifiers, when their use is applicable, are stable operation (no baseline drift) and low noise output. Most active electronic components (transistors or tubes) exhibit $1/f$ noise, that is, electronic noise increases as frequency decreases and hence is largest at very low frequencies. If we convert a low-frequency or dc signal to the chopper frequency for amplification, then electronic noise produced in the active elements in the amplifier is greatly reduced. Operation of the chopper at 1 kHz would mean a thousand-fold reduction in noise of this origin at 1 Hz.

7.4. Active Components in Preamplifiers

The basic active devices which are used in preamplifiers in chronological sequence are tubes, transistors, field-effect transistors (FET's), and chips. Chips are self-contained microelectronic semiconductor circuits which are now available in numerous configurations. Thus one can buy certain basic amplifiers and other circuits as chips for incorporation into electronic systems. Much instrument design is now being done using these ready-to-use modular units. Many chips are comparable in size to transistors and have low power requirements.

Tubes and FET's have in common the property of high input impedance. Signal-input power drain is low and output coupling is enhanced by relatively low output impedance which reduces signal loading of the amplifier stage. Tubes and FET's are generally used now for electrometer service where very high input impedances are required (see Section 7.7). FET's are used in the input circuitry of differential amplifiers (Section 7.5) and operational amplifiers (Section 7.6).

Transistors work backwards from an impedance viewpoint. They generally present a low input impedance and high output impedance which causes signal loading and problems with signal output coupling. Although fine for digital work, they are being replaced by FET's and chips for much analogue (continuous signal) work. Figure 7.19 is an abstract representation of a three-terminal active device and applies equally to the signal connections for a tube, transistor or FET.

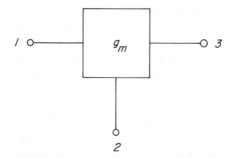

Figure 7.19. Three-terminal active device.

There are three basic circuit configurations in which a three-terminal active device can be used. Terminal 1 is the natural input terminal (grid of a vacuum tube, base of a transistor, and gate of an FET). Terminal 2 is the normal reference (cathode of a tube, emitter of a transistor, and source of an FET). Terminal 3 is the natural output (anode or plate of a tube, collector of a transistor, and drain of an FET). The configuration which produces maximum voltage amplification applies the input between 1–2 and the output is taken between 3–2. In this case, the signal is shifted in phase by 180°, or "inverted." If the input is applied between 2–1 and the output taken between 3–1, some signal amplification occurs but no phase shift. When the signal input is between 1–3 and the output between 2–3, no phase shift occurs and voltage amplification is less than unity (usually about 0.9). This configuration is used for impedance conversion: high input impedance to low output impedance. Figure 7.20 summarizes these connections and indicates the terminology being used. Methods for determining impedance levels and amplification factors, etc., using a generalized approach are outlined in an earlier publication (Ferris, 1965a).

Bias techniques and design considerations are purposely omitted here. Those interested in such matters should consult standard texts on electronics, for example, that by Brophy (1972).

7.5. Differential Amplifier

The differential amplifier is a combination circuit of two simple preamplifiers into a device which either adds or subtracts two input signals. It is represented schematically in Figure 7.21. Depending upon the internal connections in the unit, the output voltage V_o is given by

$$V_o = K(V_1 + V_2) \qquad \text{Summing connection}$$

$$\left.\begin{array}{l} V_o = K(V_1 - V_2) \\ V_o = K(V_2 - V_1) \end{array}\right\} \quad \text{Difference (usual) connection}$$

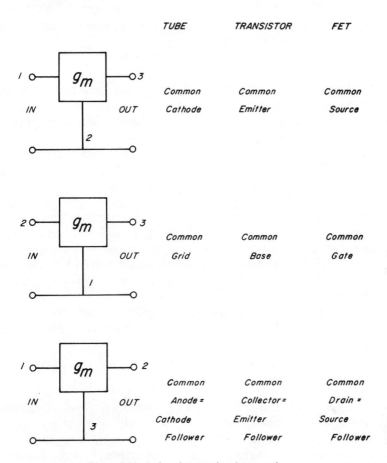

Figure 7.20. Active element signal connections.

The input voltages V_1 and V_2 are referenced to circuit ground. In the difference mode, this amplifier can be used to sense potential changes at one point in a circuit relative to another point in a circuit. A typical application is in EKG preamplifiers where two limb leads are connected to the input of a difference amplifier. They may also be used in null-indicator circuits to indicate equality

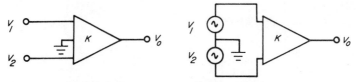

Figure 7.21. Differential amplifier.

of two voltages. When $V_1 = V_2$, then $V_o = 0$. In this case, the voltages must be equal in both magnitude and phase angle.

7.5.1. Common-Mode Signal

Let us suppose that $V_1 \neq V_2$ and that we add a voltage V_n to both. Thus, we have the augmented values $V'_1 = V_1 + V_n$, $V'_2 = V_2 + V_n$. If we use V'_1 and V'_2 as inputs to a difference amplifier, then the output will be

$$V_o = K(V'_1 - V'_2)$$

$$= K(V_1 + V_n - V_2 - V_n)$$

$$= K(V_1 - V_2)$$

The voltage V_n is said to be a common-mode voltage (signal). It is common to both input voltages and is subtracted out by the difference amplifier. The degree to which a difference amplifier is insensitive to a common-mode voltage is called the common-mode rejection ratio (CMRR). If V_n is increased by a factor of 10,000 before a factor-of-one change in V_o is noted, then the CMRR is 10,000:1.

One advantage of using difference amplifiers is that common-mode signal hum (60 Hz) and noise are eliminated by the difference-input connection as long as they do not exceed the CMRR of the amplifier.

7.6. Operational Amplifiers

Originally, operational amplifiers (op-amps) were used in computers to perform mathematical operations. With the development of semi-conductor integrated circuits, small and inexpensive op-amps have become available and are now being used in place of conventional discrete component amplifiers in many applications. Basically an op-amp is a very high-gain (theoretically infinite-gain) amplifier. In use, it is normally connected in a feedback mode of operation. That is, a portion of the output voltage is

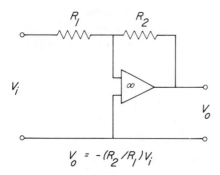

$$V_o = -(R_2/R_1)V_i$$

Figure 7.22. Basic operational amplifier circuit.

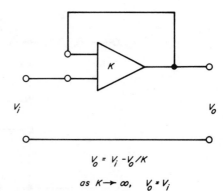

Figure 7.23. Operational amplifier voltage
follower.

$$V_o = V_i - V_o / K$$

$$\text{as } K \to \infty, \quad V_o = V_i$$

returned to the input terminals with 180° of phase shift such that the returned
voltage subtracts from the input voltage. In this manner, amplifier gain is
both controlled and stabilized. The basic network configuration is shown in
Figure 7.22.

The advantage of op-amps is that they can be used as building blocks
to construct a number of useful circuits. In most cases, only resistors are
needed to complete a design. We will now examine some particular circuits.
At this point, the actual arrangement of active devices to make an op-amp
is of no concern to us. We will simply use the device as it comes from the
manufacturer.

The op-amp connection to make a follower circuit (emitter follower,
etc.) is shown in Figure 7.23.

Another useful circuit is the noninverting amplifier which provides gain
without phase shift. Circuit connections using an op-amp are shown in
Figure 7.24.

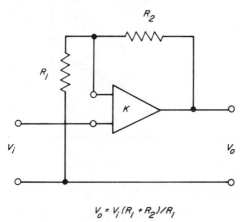

$$V_o = V_i (R_1 + R_2) / R_1$$

Figure 7.24. Noninverting operational amplifier.

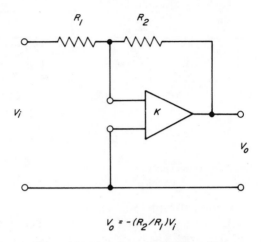

$$V_0 = -(R_2/R_1)V_i$$

Figure 7.25. Inverting operational amplifier.

If a 180° phase shift is desired (inverting amplifier), then the signal connections are as shown in Figure 7.25.

A single op-amp may be used as a differential amplifier by connecting it as shown in Figure 7.26.

In this section, we have discussed signal connections only for operational amplifiers. It is also necessary to provide dc bias voltages (see Figure 2.15, for example). We will not discuss biasing here. Normally this information is

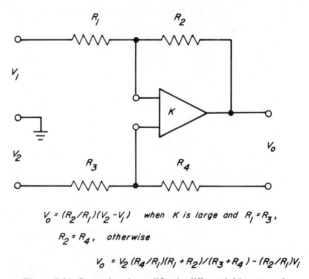

$$V_0 = (R_2/R_1)(V_2 - V_1) \quad \text{when } K \text{ is large and } R_1 = R_3,$$

$$R_2 = R_4, \quad \text{otherwise}$$

$$V_0 = V_2 (R_4/R_1)(R_1 + R_2)/(R_3 + R_4) - (R_2/R_1)V_1$$

Figure 7.26. Operational amplifier in differential input mode.

supplied in detail on the manufacturer's specification sheets that are provided with op-amps when purchased.

The connections which we have shown are for dc input. If ac operation is desired, all that is necessary is the insertion of a coupling capacitor in series with the input signal lead.

We have not discussed values of resistance to be used with op-amp circuits. In most cases, the manufacturer will indicate minimum and maximum values. If we examine the voltage gain expression for the inverting amplifier,

$$V_o = -(R_2/R_1)V_i$$

we see that we can adjust gain simply by selecting R_1 and R_2 in the correct ratio. Thus if we require an amplifier with a gain of 100, then $(R_2/R_1) = 100$ or $R_2 = 100R_1$. For $R_1 = 1\,k\Omega$, $R_2 = 100\,k\Omega$.

There are various factors in addition to biasing which must be considered when using op-amps. Op-amps will not function above a predetermined value for V_o. When V_o exceeds the stated (specification sheet) value, saturation occurs. Since op-amps can be operated in both inverting and non-inverting modes, saturation is normally symmetric and is determined when $V_o = \pm V_{sat}$.

Typical gain values for open-loop gain are $> 100,000$, defined for the basic op-amp without any feedback connections (return connections, either direct or through resistors, from output to input). Closed-loop values for gain are determined from the expression relating V_o and V_i in Figures 22–26, and depend upon resistance ratios.

Op-amp bandwith and compensation must be considered. Generally op-amps are low-frequency devices. Unless an input dc-blocking capacitor is used in the signal path, they respond to 0 frequency (direct current). The high-frequency 3-db point is usually in the range 1–10 kHz. Bandwidth depends upon how much gain is desired. The higher the gain, of course, the lower the BW. A typical op-amp characteristic is plotted in Figure 7.27. From the figure, we note that for a gain of 1000, we can get a bandwidth of about 1.1 kHz. If the gain is reduced to 10, the bandwidth is greater than 100 kHz.

Op-amps generally contain compensation networks to increase stability and external bandwidth. In some cases, these are adjustable. Again, the manufacturer's specification sheets provide this information. Other factors to be considered in using op-amps are noise, drift in operating point, and temperature stability. These vary with different op-amp configurations. Good general references for these problems are handbooks supplied by manufacturers of op-amps (Burr–Brown, 1963, for example).

Op-amps function quite well in performing the operations of integration and differentiation. Basic circuits are shown in Figure 7.28.

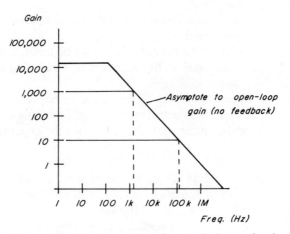

Figure 7.27. Gain characteristic for a typical operational amplifier.

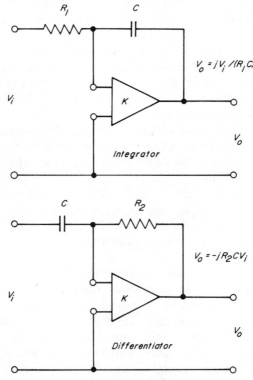

Figure 7.28. Operational amplifier integrator and differentiator circuits.

7.7. Electrometer Preamplifiers

Electrometer preamplifiers are used where very low current drain from a signal source is important. Input impedances of such circuits lie in the range from 10^9 to $10^{14}\,\Omega$. They are used as preamplifiers for microelectrode recording, and null sensing circuits in potentiometric devices such as pH meters. The active elements in these circuits are either electrometer vacuum-tubes or insulated-gate FET's. Ordinary transistors are inapplicable because of their low input impedance.

When vacuum devices are used, either tubes designed especially for electrometer service can be employed, or certain miniature pentode tubes can be connected for electrometer operation. Normally electrometers are used as dc signal amplifiers when very small currents are available. Because of their high input impedance, their time constants are long and they don't generally respond rapidly to fast transients nor do they recover quickly from signal overload. Vacuum electrometers are sensitive to a number of environmental factors and require much care in use. FET's are more rugged in this respect, although one must be careful that excessive charge buildup does not occur on the gate terminal, as this can destroy an insulated-gate FET.

Electrometer circuits *per se* are not directly applicable to physiological recording systems because of their long time constants. Even if the shunt capacitance of the input circuit is only 1 pF, for the range of input shunt resistance from 10^9 to $10^{14}\,\Omega$, the time constants of the input circuit will vary from 1 msec to 100 sec. These time constants are excessive. Where electrometer techniques do find application are in special feedback circuits designed to cancel input capacitance. These are discussed below.

7.7.1. Negative-Input-Capacitance Preamplifier

In 1949, Bell published a circuit for use with certain video amplifier systems. The system is an adjustable self-neutralizing feedback amplifier, which when properly adjusted, has no shunt input capacitance. A functional diagram is shown in Figure 7.29. The input capacitance of the system shown in Figure 7.29a is

$$C_{in} = (K - 1)C_f \qquad \text{(Bell, 1949)}$$

where K is the open-loop gain (without C_f connected) of the amplifier and C_f is the value of the feedback capacitance. In Figure 7.29b, the input capacitance C_i and resistance R_i are shown explicitly. Theoretically, it is possible to make

$$C_{in} = -C_i$$

by proper adjustment of K and C_f. Namely, if

$$K < 1$$

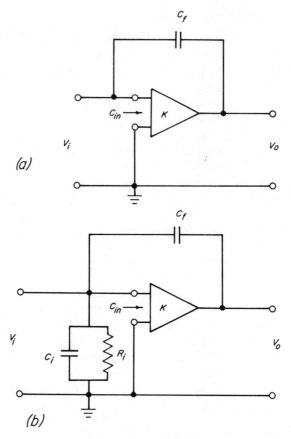

Figure 7.29. Functional diagram for negative-input-capacity amplifier.

and C_f is positive such that

$$(K - 1)C_f = -C_i$$

the condition

$$C_{\text{in}} = -C_i$$

can be satisfied. This condition has been described in some detail with regard to operational amplifier and integrator circuits (Millman and Taub, 1956).

Several practical designs have evolved from the functional representation shown in Figure 7.29. These have been described by Amatniek (1958) and Bak (1958). Generalized schematic diagrams are presented in Figure 7.30.

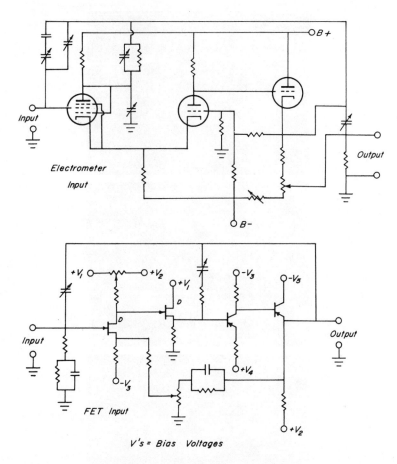

Figure 7.30. Negative-capacity-input amplifiers (basic circuits) (based upon Amatniek, 1958, and Bioelectric Instruments, Inc., circuit diagram for NF 1 amplifier).

Electrometer tube input circuits are preferable to transistor input, although field-effect transistors are almost as good. Input resistances of the order of $10^{12}\ \Omega$ are possible with tubes and of the order of $10^{11}\ \Omega$ with FET's in the feedback configurations indicated.

A modified form of the Bak unit has been used by the author at the University of Maryland and performance curves are shown in Figure 7.31. The effect of the setting of the neutralizing capacitor upon transient response and signal wave shape is shown.

In the design of negative-input-capacity amplifiers, one must insure that tube grid current or gate signal current is minimized. These currents

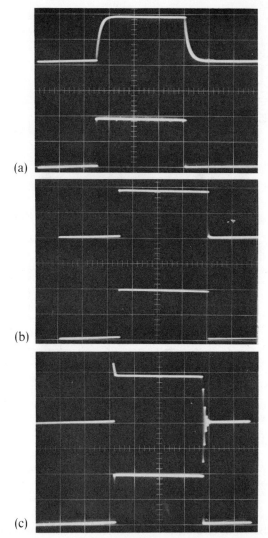

Figure 7.31. Transient response for modified Bak Electrometer amplifier. Signal level: Approximately 5 V peak-to-peak.

(a) Effect of input capacity upon square wave input. The input signal (lower trace) is applied through a 20-MΩ resistor with associated cable capacity. The upper trace is the output from the electrometer adjusted for unity gain but without compensation. Time base: 2 msec/cm.

(b) Upper trace: output from electrometer under unity gain condition with feedback circuit adjusted to neutralize the input capacitance. Lower trace: same as in (a).

(c) Upper trace: output from electrometer under unity gain condition with feedback circuit adjusted for overcompensation. Note the beginning oscillations. Lower trace: same as in (a).

can have a deleterious effect upon fidelity of measurement by modifying the properties of the physiological system (such as a single cell), causing baseline drift, polarizing the electrodes, causing a nonlinear variation in amplifier input resistance and concomitant change in signal amplitude, which generates noise. A safe grid current with regard to these considerations is one that is less than 10^{-12} A. The use of fluid-bridge electrodes lowers the probability that the electrodes will be polarized. A higher probability of electrode polarization exists when metal electrodes come into direct contact with the physiological system.

7.8. The Voltage-Clamp Circuit and Feedback

The voltage-clamp technique was introduced by Cole in 1949. A general survey of the principle of operation and various shortcomings is presented by Moore and Cole (1963). A voltage-clamp circuit provides a means for holding constant the potential difference across the resting or active plasma membrane of a neuron (axon), within the limitations of practical electrodes. The discussion here is concerned primarily with the amplifier system being used.

Figure 7.32 illustrates the basic clamp circuit. A differential amplifier is used. The mathematical relations which describe voltage-clamp operation are

$$K(V_i - V_m) = V_o$$

$$V_o = RI + V_m$$

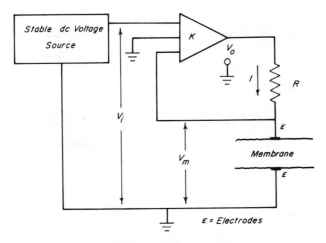

Figure 7.32. Basic voltage-clamp system.

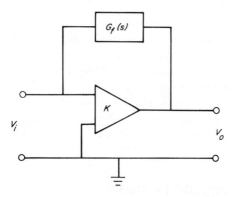

Figure 7.33. Basic feedback amplifier.

where I is the current delivered by the amplifier to maintain the voltage V_m across the membrane. Substituting,

$$K(V_i - V_m) = RI + V_m$$

$$V_m = \frac{KV_i}{1 + K} - \frac{RI}{1 + K}$$

In order to find an approximate value for K, we examine the "error" signal applied to the input of the difference amplifier. If E is the error, then

$$E = V_i - V_m$$

$$= V_i - \frac{KV_i}{1 + K} + \frac{RI}{1 + K}$$

$$E = \frac{V_i + RI}{1 + K}$$

If E is to be small, K must be large. Normally a differential op-amp would be used with $K > 10,000$.

7.9. Feedback

The circuits described in Sections 7.6–7.8 have been dependent upon feedback. Amplifiers have been described in which part of the output signal was returned to the input. Feedback is used to stabilize amplifier operation or to give a controlled output. Figure 7.33 illustrates a basic feedback amplifier. The output–input relation is

$$V_0 = KV_i + KG_f(s)V_o$$

$$V_o = \frac{K}{1 - KG_f(s)}V_i$$

$G_f(s)$ may be a simple resistor network or it may be a frequency-selective network. If $G_f(s)$ is simply a resistor network, the effect of the feedback will be to reduce the overall gain of the amplifier, since

$$\text{Amplifier gain} = \frac{V_o}{V_i} = \frac{K}{1 - KG_f(s)}$$

For a resistive network, $G_f(s)$ is a number which may be greater or less than unity. The feedback in this case produces gain stabilization. Amplifier frequency response is increased (gain–bandwidth relation) and stability is improved. There is also a reduction in electrical noise at the output because of signal cancellation.

When $G_f(s)$ is frequency-selective, then phase shift occurs in the signal which passes through the feedback network. If the return signal is out of phase with V_i, then signal subtraction occurs and the feedback is said to be negative. If the return signal is in phase with V_i, then positive feedback occurs. Negative feedback leads to stabilized operation and positive feedback to unstable operation. Positive feedback through a very frequency-selective network provides the basis for tuned (frequency-selective) amplifiers and oscillators. Positive feedback increases selectivity (decrease in BW), increases gain and promotes instability. Negative feedback decreases selectivity (increase in BW), decreases gain and promotes stability.

Feedback is also used to provide a constant (controlled) voltage as shown in Figure 7.34. V_o is the controlled voltage. The output is sampled by the voltage-divider network of R_1 and R_2 and the feedback voltage V_f is compared against a reference V_i.

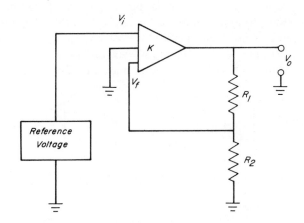

Figure 7.34. Controlled-output feedback system.

$$V_o = K(V_i - V_f)$$

$$V_f = \frac{R_2}{R_1 + R_2} V_o$$

$$V_o = \frac{K(R_1 + R_2)}{R_1 + R_2(1 + K)} V_i$$

Initially V_i and the ratio $R_2/(R_1 + R_2)$ are adjusted to give a particular V_o. If V_o then starts to change, V_f changes such that the amplifier input $(V_i - V_f)$ changes, which then returns V_o to its preset value. The error signal applied to the amplifier is

$$E = V_i - V_f$$

$$= \frac{R_1 + R_2(1 + K)}{K(R_1 + R_2)} V_o - \frac{R_2}{R_1 + R_2} V_o$$

$$E = V_o/K$$

Thus the error E is minimized if K is large. For example, let $V_o = 100\,\text{mV}$ (typical value in cellular recording) and let the permissible error be 0.1, then $K = 1000$.

A full treatment of feedback systems is beyond the scope of this text and the reader is referred to the literature (see Millsum, 1966, for example).

7.10. Isolation Networks

In excitation–response systems such as are used in neurophysiological studies and some impedance plethysmography techniques, a stimulus artifact can result from the stimulating current reaching the recording electrodes directly. In some instances, this can be desirable when propagation-time measurements are being made. In most physiological work, stimulus artifact is undesirable. Artifact signals are generated when both the stimulating and recording systems have a common ground. To avoid this situation,

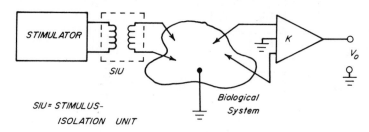

Figure 7.35. Use of stimulus-isolation unit.

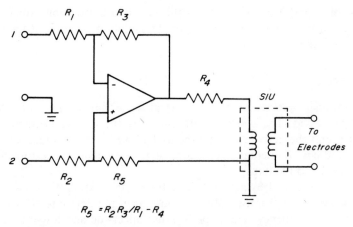

$$R_5 = R_2 R_3 / R_1 - R_4$$

Figure 7.36. Constant current source. A negative pulse may be applied at input 1 and a positive pulse at input 2 to synthesize a complex pulse. Either terminal may be grounded if only one input pulse is required.

the stimulator is isolated from the ground system of the recording network by an isolation network which generally contains a pulse transformer as shown in Figure 7.35.

In this system, a differential amplifier is used and its reference ground is the same as the ground reference in the physiological system being studied. The stimulus unit (pulse generator) is floating with respect to the rest of the system.

Generally speaking, the most stable operation of stimulus–response systems occurs when constant current signal sources are used. These can be developed through the use of an op-amp as shown in Figure 7.36.

In some instances, a more sophisticated stimulus-isolation unit is used as is shown in block-diagram form in Figure 7.37 (Strong, 1970).

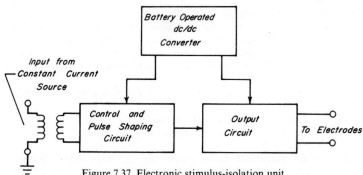

Figure 7.37. Electronic stimulus-isolation unit.

Typical performance values for this system are: isolation from constant current source, 5 pF coupling at $> 10^{10}\ \Omega$; isolation of output from ground, 30 pF coupling at $> 10^{10}\ \Omega$.

7.11. Noise

Amplifier systems generate electrical noise which adds to the signals being processed. The source of this noise is the passive elements in the circuitry and the active elements themselves. Other noise sources are electrodes and environmental electrical noise from switchgear, motors, fluorescent lamps, and the like.

Noise produced by active devices (tubes, transistors, FET's) is usually $1/f$ type; that is, the noise voltage produced in the device is inversely proportional to frequency. This type of noise affects dc and low-frequency ac amplifiers in particular.

Noise produced in passive circuit elements (resistors especially) is usually of thermal origin and has the mean-square voltage value

$$v^2 = 4kTR\,\Delta f$$

where k is Boltzmann's constant, T is Kelvin temperature, R is the equivalent resistance of the noise generating network, and Δf is the bandwidth of the instrument used to measure v^2. Thermal noise is most apparent in wideband (ac) amplifier systems.

Signal-from-noise extraction and filtering techniques are discussed in Chapter 9.

7.12. References

Amatniek, E., 1958, Measurement of bioelectric potentials with microelectrodes and neutralized input capacity amplifiers, *Trans. I.R.E., PGME* **10**:3.

Arguimbau, L. B. and Adler, R. B., 1956, *Vacuum-tube Circuits and Transistors*, John Wiley and Sons, New York.

Bak, A. F., 1958, A unity gain amplifier, *EEG Clin. Neurophys.* **10**:745.

Bell, P. R., 1949, Negative capacity amplifier, Appendix A, pp. 767–770, in *Waveforms* (B. R. Chance *et al.*, eds.), *MIT Rad. Lab. Series*, McGraw-Hill, New York.

Brophy, J. J., 1972, *Basic Electronics for Scientists, 2nd Ed.*, McGraw-Hill, New York.

Burr–Brown, 1963, *Handbook of Operational Amplifiers*, Burr–Brown Research Corp., Tucson, Arizona.

Cole, K. S., 1949, Dynamic electrical characteristics of the squid axon membrane, *Arch. Sci. Physiol.* **3**:253.

Ferris, C. D., 1962, *Linear Network Theory*, C. E. Merrill, Columbus.

Ferris, C. D., 1965a, A general network representation for three-terminal active devices, *Trans. IEEE, E* **8**(4):119.

Ferris, C. D., 1965b, Low frequency bridge detector, *Rev. Sci. Instr.* **36**(11):1652.

Millman, J. and Taub, H., 1956, *Pulse and Digital Circuits*, McGraw-Hill, New York.

Millsum, J. H., 1966, *Biological Control Systems Analysis*, McGraw-Hill, New York.

Moore, J. W. and Cole, K. S., 1963, Voltage clamp techniques, in *Physical Techniques in Biological Research, Vol. 6* (W. L. Nastuk, ed.), Academic Press, New York.

Strong, P., 1970, *Biophysical Measurements*, Tektronix Corporation, Beaverton, Oregon.

CHAPTER 8

Specialized Electrodes

There are a number of different types of electrodes which are used for specialized applications. All of them depend upon the general principles outlined in Chapters 1–6. They are sufficiently specialized in application, however, that separate discussion is deemed advisable. Further discussion, of some historical interest, is presented by Geddes and Baker (1968) in Chapter 11 and by Geddes (1972) in Chapters 2–4.

8.1. Body-Cavity Electrodes

The natural body openings lend themselves to electrode placement for diagnostic (recording) and therapeutic (stimulating) use. The areas normally considered are the esophagus, trachea, and vagina. The genitourinary areas are commonly sites for active electrode use in various surgical techniques, specifically use of electrocauteries and electric scalpels. These sorts of electrodes are not within the scope of material treated in this text.

Endotracheal and endoesophageal electrodes have been used in cardiac studies (Ferris, 1966) both for recording and stimulation. The general form of the electrode is shown in Figure 1.2. For canine studies, an endotracheal electrode can be fabricated from a No. 36 (8.5-mm) endotracheal tube. The inflatable cuff on the distal end of the tube is coated internally and externally with flexible, conducting silver paint. The paint is also poured inside the tube over its entire length. A wire connection is made to the paint coating to form the electrical connection to the electrode. In use, the pressure cuff is inflated to assure positive contact between the external silver paint and the inner wall of the trachea.

The advantage of an electrode of this sort is that it can be used during surgical procedures in conjunction with an anesthesia machine, as long as the subject is intubulated with a tracheal tube. If the electrode is used in conjunction with an external electrode on the chest surface, EKG records can be recorded and ventricular fibrillation can be converted, should it occur, by connecting the electrodes to a defibrillator. Less power is required for defibrillation with this electrode configuration than with conventional chest surface electrodes. Defibrillation should not be attempted if patient's lungs contain an explosive anesthetic.

The esophagus may be used in the same manner as the trachea for electrode placement. Again for canines, a No. 26 endotracheal tube is used. Instead of using silver paint, a three-inch braided-copper-wire-mesh bulb is attached to the tube tip. A wire stylette is inserted coaxially in the tube and attached to the bulb. In use, the electrode is inserted into the esophagus, keeping the bulb extended in the longitudinal direction. After placement, the stylette is pulled, which causes the mesh to contract, thus making firm contact with the inner wall of the esophagus. The wire stylette forms the electrical connection to the electrode.

Vaginal electrodes have been used in studies of reproductive function by Burr *et al.* (1935), Rock *et al.* (1937), and Parsons *et al.* (1958). Techniques reported to date are rather crude and the subject must be essentially immobile. It is suggested that a conforming cylinder or pessary electrode be used in conjunction with a radio telemetry system. Such a system for studies of ovarian function is being developed by Ferris and Weeks (University of Wyoming). With this system, the subject will have complete mobility.

8.2. Contoured Electrodes and Specialized Cardiac Electrodes

The electrodes described in this section were developed primarily for cardiac studies related to ventricular fibrillation and defibrillation. These configurations, however, are applicable to other studies.

8.2.1. Flexible Electrodes

Occasionally it is desirable to have an electrode which can conform to the body surface or to the surface shape of an organ. There are several ways of approaching this problem, two of which are presented here. Large-surface-area electrodes (> 2 in.) can be fabricated from a commercially available stainless steel mesh material (light weight) similar to chain mail. This flexible material is backed with a one-half-inch layer of foam rubber, which in turn can be glued to a solid matrix. Either wood or a plastic material can be used for the matrix. The advantage of the rigid matrix is that handles can be attached so that these electrodes can be held firmly in place in such applications as closed-chest ventricular defibrillation.

Contact wires are welded to the reverse side of the mesh (metal–foam interface side) for electrical connection. The metal mesh can be bonded, with appropriate adhesives, directly to the foam material. If a smooth electrode surface is required, the mesh, in turn, may be covered with a smooth sheet of flexible metal foil such as aluminum foil used for cooking purposes. This is bonded to the mesh by crimping at the electrode edges. The mesh supplies mechanical strength which the foil alone would not possess. The rigid matrix backing is not essential and can be omitted if the electrode is held in place by straps or tape.

A design which we have found successful when chronic electrodes are placed on organ surfaces is the crocheted electrode. It is made by crocheting fine flexible wire, such as stainless-steel alloy surgical sutures (Ethi-Pack® size 00, B & S 28-gauge multifilament surgical sutures), into a circular disc of the required diameter (usually 1–2 cm). A long tail of wire is left attached to the crocheted electrode to form the connecting lead. This lead is insulated either by coating it with medical-grade silastic rubber, or by fitting it through a fine plastic catheter tube. Electrodes fabricated in this way are attached to the organ under study by placing silk sutures through the holes in the crocheted mesh. We have found such electrodes extremely useful in cardiac studies, as the electrodes conform to the ventricular surface and can be firmly attached.

Flexible electrodes (both mesh and crocheted) are useful in that they adapt to the contour of an organ or body surface and reduce the problems associated with variations in contact generally experienced with rigid electrodes.

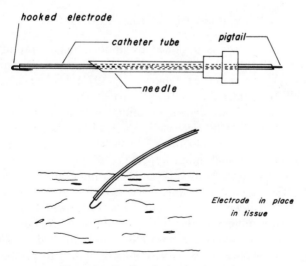

Figure 8.1. Cardiac single-wire electrode.

8.2.2. Other Chronic Electrodes

When chronic flexible-needlelike electrodes are required, a useful technique is illustrated by the electrode shown in Figure 8.1 and in Figure 1.1. Originally this electrode was proposed for cardiac emergency treatment such as in Adams–Stokes syndrome (Ferris and Cowley, 1968). The electrode is an insulated, coaxial, bipolar arrangement using a No. 18 hypodermic needle, 6 in. in length (in this case), with an internal flexible stylette of smaller diameter. The electrode can be inserted through the body surface into an organ. When the hollow needle is carefully withdrawn, the flexible stylette hooks into place, anchoring the electrode tip, and leaves a flexible plastic tube in place of the rigid needle. To withdraw the electrode, the rigid needle is reinserted coaxially with the electrode assembly, and the complete assembly is withdrawn. Methods for placing single wires are illustrated in the figure.

8.3. Anodized Electrodes

A problem exists in chronic recording of body surface potentials, such as EKG recording from individuals in space flights. Conventional metal surface electrodes require the use of some sort of wetting agent to maintain electrical contact. In the chronic situation, this agent dries out and contact problems develop. Skin disturbances and irritation may also occur. In an attempt to solve this problem, Richardson et al. (1968) proposed an insulated or capacitive electrode based on the following principle:

Body surface (skin)|Dielectric|Metal|FET

The dielectric layer is obtained by anodization to produce a metal oxide film on the metal substrate. Both aluminum and tantalum foils (99.99 % purity) have been used (Lagow et al., 1971). The gate of a field-effect transistor (FET) is soldered directly to the back of the electrode. In this manner, noise and extraneous signals are minimized. As shown in Figure 8.2, the entire assembly can be electrically shielded and potted in epoxy. The electrical circuit is shown in Figure 8.3. A simple source-follower configuration is used. The very high input impedance of the FET is utilized for the capacitive coupling to the body surface. The output from the source follower, after passing through a remote unit, is fed directly into a recorder.

It has been shown experimentally (McMullin and Pryor, 1961) that aluminum oxide is quite porous and in film form is attacked by chloride ions. Thus its use in electrodes of this nature is contraindicated as skin moisture containing NaCl would work its way into the pores to produce chloride corrosion. Electrical noise and electrode instability occur as a consequence

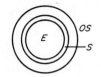

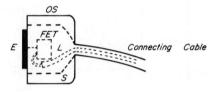

E = anodized electrode   S = copper shield

FET = field-effect transistor L = FET connections into cable

OS = molded epoxy outer shell

Figure 8.2. Anodized electrode.

of this action. On the other hand, tantalum forms dense and nonporous films which are apparently chloride resistant.

Anodization is accomplished by cleaning the tantalum metal in a detergent and degreasing in boiling trichloroethylene. A technique used in the vacuum-tube industry involves placing the tantalum in a reflux condensor and distilling the degreasing solvent over the metal. The tantalum

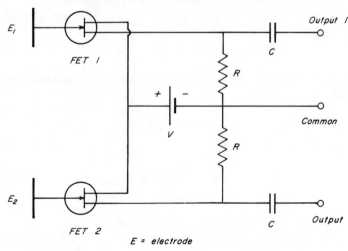

Figure 8.3. Circuit diagram for two-electrode-pair system using anodized electrodes. The shields and their connections are not shown.

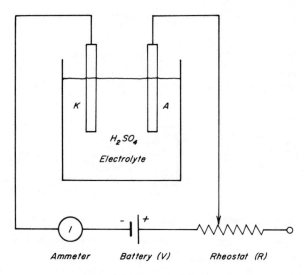

Ammeter   Battery (V)   Rheostat (R)

Tantalum cathode (K) & anode (A)

Figure 8.4. Anodizing circuit.

parts are then removed to a hydrogen-atmosphere electric furnace to outgas the metal and remove surface impurities. For the anodizing electrolyte, 0.1 molar H_2SO_4 is used, at a constant current density of 0.1 mA/cm^2 (Lagow et al., 1971). The resistor shown in Figure 8.4 must be adjusted periodically, as the oxide film builds up, to maintain this current density.

As shown by Young (1961), the oxide thickness T in angstroms is given by

$$T = 17.5\ V_f$$

where V_f is the forming voltage. Thus if V_f is 10 V, one can expect an oxide layer thickness of 175 Å. Electrode capacitance is given by

$$C = \varepsilon_0 \varepsilon A / T$$

where

$$\varepsilon_0 = 8.85 \times 10^{-14}\ \text{F/cm}$$

$$\varepsilon = 27.6 \text{ for tantalum}$$

$$A = \text{electrode surface area in cm}^2$$

$$T = 17.5\ V_f\ \text{Å}$$

$$= 17.5\ V_f \times 10^{-8}\ \text{cm}$$

Thus

$$C/A = \varepsilon_0 \varepsilon /(17.5\, V_f)$$
$$= 14\, \mu\text{F}/\text{cm}^2$$

Impedance bridge techniques can be used to establish the electrical properties of the oxide film. The resistivity ρ of the oxide layer is determined by measuring the dissipation factor of the capacitance. Reported values for ρ are of the order of 10^9 Ω-cm (Lagow *et al.*, 1971).

Ta_2O_5 electrodes appear to be quite stable in chronic recording situations.

8.4. Self-Wetting Electrodes

Another approach to chronic recording from the body surface is the use of self-wetting electrodes. For this application, a hygroscopic material is impregnated in the electrode surface. The electrode must be porous so that it can retain the chemical, and the chemical in a wetted state must have a low electrical resistivity to promote good electrical contact. A proposed design is shown in Figure 8.5. A stainless-steel backing plate is used. This is faced by a thin layer of balsa wood impregnated with lithium fluoride (Morris, 1967).

Three major problems with this approach are skin irritation, interface problems at the wood–metal interface, and varying electrode resistance as a function of moisture content in the balsa layer.

8.5. Suction Electrode

A specialized microelectrode-type electrode was developed by Florey and Kriebel (1966) for use in certain types of neurological recording. The structure is basically a suction pipette as shown in Figure 8.6a. The tip of the electrode is made slightly smaller in internal diameter than the diameter of the axon or nerve bundle being studied. The electrode assembly is filled with isotonic solution and suction applied to the nerve end until the desired

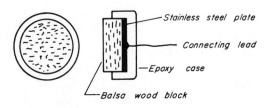

Figure 8.5. Self-wetting (balsa) electrode.

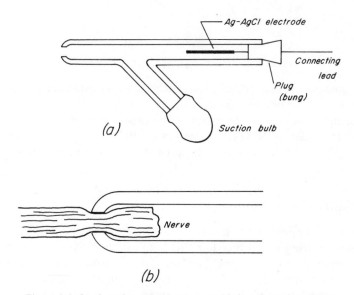

Figure 8.6. Suction electrode. A more sophisticated version uses a silver–silver chloride button electrode imbedded in the wall of a plastic electrode holder. Only the pore tip is glass. Suction is obtained via a thin catheter tube attached to a standard hypodermic syringe.

amount of cell material is in the electrode lumen. A certain amount of pinching down of the tissue occurs as shown in Figure 8.6b. This is desirable to insure a tight seal at the junction between the internal fluid bridge in the electrode and the external fluid which bathes the preparation. Electrical connection to the preparation is achieved by means of a fluid salt bridge and a silver–silver chloride electrode. An indifferent electrode is used externally to the suction electrode to complete the connection to the recording system.

This type of electrode provides positive contact with a nerve fiber for long-term stimulus-response recording (see also Section 10.6).

8.6. Percutaneous Electrodes

Some research has been conducted on the types of electrodes to use in chronic recording and stimulating involved with electromyographic work and myoelectric control problems. Two rather different types of electrode systems have been proposed (Caldwell and Reswick, 1967; Miller and Brooks, 1971). These electrodes are shown in Figure 8.7. Wire implants have been used by Schane (1973).

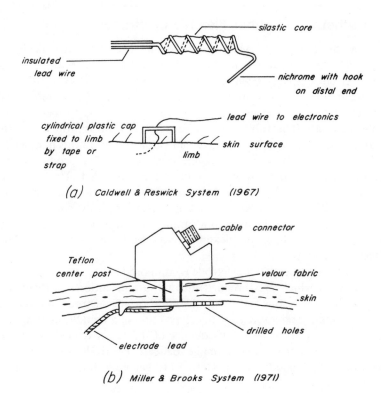

Figure 8.7. Percutaneous electrode systems.

The Caldwell–Reswick electrode is a helix of nichrome resistance wire filled internally with a silicone rubber core. A retaining hook is located on the distal end. The electrode is placed by means of a 23-gauge hypodermic needle in the same manner as the cardiac electrode shown in Figure 8.1. This electrode can be placed directly into a muscle. The silastic core and the helical assembly minimize electrode breakage. Placement and removal of the electrode is quick and easy.

As shown in Figure 8.7, the Miller–Brooks system is rather complicated. The electrode basic assembly is Teflon fitted with appropriate metal connectors. The general principle is that of a snap fastener with skin rather than fabric as the associated medium. The center post of the assembly is wrapped with velour fabric which is bonded to the post by adhesive. The purpose of this is to form a substrate to which tissue can adhere after surgical implantation. While the developers of this scheme have reported excellent results, placement and removal of the "snap" assembly is cumbersome.

8.7. Catheter and Other Flexible Electrodes

Some discussion of flexible electrodes has been presented in Sections 8.1 and 8.2. Another useful method, applicable to both body-cavity and catheter electrodes, has been described by Schaudinischky *et al.* (1968). Their technique involves sputtering of thin metallic films onto flexible substrate materials such as sheet latex rubber, silicone rubber, or other flexible materials. The substrate can be masked with appropriate cutouts so that any desired electrode patterns can be produced. In this manner both the electrodes and the necessary connecting leads can be produced simultaneously. Vacuum deposition of gold is recommended. The connecting leads can be insulated by further deposition of a nonconducting metal oxide, or painting with a flexible nonconducting rubber or plastic compound.

This technique can be used to form metallic electrodes on catheter tips and balloons as well as on flexible sheets which can contour to body surfaces.

The more usual catheter electrode is composed of a single wire or several insulated wires which are threaded through a conventional catheter tube. The tip of the tube is thus fitted with the appropriate electrode or electrodes.

Sputtering is preferable to vapor deposition of a metallic coating. High temperatures can result as a consequence of the vapor deposition process. Excessive temperatures may damage delicate electrode substrates. The temperatures associated with sputtering are generally low enough that no damage occurs.

8.8. Spray-On Electrodes

A flexible electrode has been developed for aerospace use which is sprayed directly on the subject (Roman, 1966; Patten *et al.*, 1966). It is suitable for electrocardiographic recording. The skin surface is first prepared by brushing electrode jelly into the skin with a stiff brush. Excess jelly is wiped off with dry gauze. The electrode is formed by painting or spraying conductive adhesive onto the prepared area, to produce a 2-cm-diameter electrode. Electrical connection is made by placing a silver-plated copper wire into the drying adhesive. Upon drying, the wire is held captive, thus forming a lead. When thoroughly dry, the entire assembly is covered with insulating cement.

The formula for the conducting adhesive is as follows: 43 g Duco cement (DuPont S/N 6241), 23 g silver power (Handy and Harman Silflake 135), 125 cc acetone.

Spray-on electrodes are useful for chronic recording and motional artifacts are minimized.

8.9. Vapor-Deposited Electrodes

For substrates that can withstand heat, vapor deposition of metallic coatings produces excellent results. Thin films can be obtained, if required, and surface thickness can be controlled if care is taken. Metallic-coated microelectrodes can be made by this process. Glass micropipettes can be coated with a variety of metals. Either hollow or solid glass stock may be used to fabricate the substrate glass electrode. Such electrodes, with platinum coating, are available from the Transidyne General Corporation, Ann Arbor, Michigan, under the trade name Microtrodes. Fabrication techniques have been described by Brown and McCusker (1968).

Integrated circuit techniques and processes have also been used to form small electrodes (Wise and Starr, 1969). These are still in the experimental stage.

8.10. Miscellaneous Electrodes

A wide variety of electrodes have been reported for special purposes. It is hard to categorize them. Some of these miscellaneous types are cutting electrodes, depth electrodes, and recessed electrodes.

Cutting electrodes have been fabricated from a large variety of available items including safety pins, surgical cutting needles, and electronic alligator or battery clips. Generally their use is in animal experimentation where there are problems of surface hair and tough hides. These are generally "make do" electrodes when one is not concerned with trauma to the subject (see Geddes, 1972, Chapter 3). Depending upon electrode configuration, cutting electrodes may be used for either recording or stimulating.

Depth electrodes are usually used for chronic recording of signals from the brain. They generally consist of a bundle of insulated electrodes, which may be arranged so that the electrode tips penetrate to different depths. Stainless-steel or platinum wires insulated by Teflon sleeves are frequently used. The insulated wires may be bonded together in bundles up to several dozen wires. The bundle thus forms a massive probe electrode which is capable of extracting signals from various depths simultaneously. Delgado (1964) and Ray *et al.* (1965) discuss in detail the fabrication and use of these electrodes.

Many electrodes qualify as recessed electrodes. A metal recording electrode linked by a salt bridge to a potential source is a recessed electrode. Recessed skin-surface electrodes are generally cup-shaped. The gap between the skin surface and the metallic cup electrode is filled with electrode paste or some other electrolyte to maintain good electrical contact between the electrode and the surface which it contacts. The advantage of this type of electrode is a constant and uniform electrical contact with some corresponding surface. The disadvantage of some recessed electrodes is that they are hard to apply.

Some recessed electrodes have suction bulbs attached to facilitate application to a body surface. This type is used in EKG recording from the chest surface.

Recessed Ag–AgCl pellet electrodes are available for EKG and especially EEG recording.

Frequently recessed electrodes are circular in cross-section and are equipped with a rubber gasket ring which forms a tight seal between the electrode assembly and the surface which the electrode contacts.

8.11. Electrode-Anchoring Techniques

Various techniques have been discussed for anchoring electrodes as this text has developed. Internal electrodes are generally held in position by sutures, catheters, inflatable cuffs, or other methods such as described in Section 8.6. Hagfors and Schwartz (1966) have described a novel technique for anchoring electrodes used for chronic neural stimulation. In essence, it is a small nerve chamber.

A major problem is the anchoring of skin-surface electrodes to maintain reasonably constant contact-impedance and to prevent motion artifacts. Usual techniques involve elastic bands and adhesive tape. Conventional metal electrodes generally require the use of an electrode paste to maintain contact and reduce motional problems. Pasteless electrodes as described in Section 8.3 avoid most of the problems associated with conventional metallic electrodes, but they still must be held in place by a strap or belt.

Recording from recumbent subjects is not nearly so much of a problem as recording from moving subjects. With exercising subjects, motional artifacts can become extreme. Pasteless electrodes held in place by flexible elastic, but nonconfining, straps appear to be the solution. Spray-on electrodes have also been used. Amatneek and Simenhoff (1966) have described a method which is reasonably successful. The male half of a standard garment snap is attached to two layers of micropore adhesive tape. If additional surface area is required, a metal screen may be attached as shown in Figure 8.8. Stainless-steel surgical mesh is a satisfactory material. The electrode is simply taped in place on the skin, after the electrode site has been mildly abraded. Electrode paste may also be used. Connection to the electrode is achieved by soldering a lead to the female half of the snap and then snapping it to the male half. The ultimate solution to the problem of skin-surface electrodes may well be noncontact sensors.

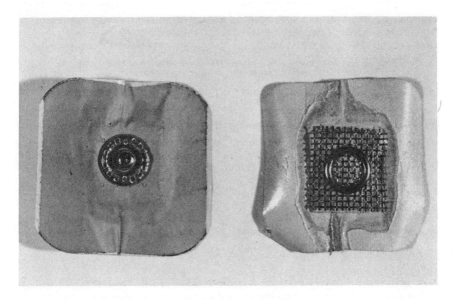

Figure 8.8. Snap electrode.

8.12. References*

Amatneek, K. F. and Simenhoff, M. L., 1966, A 24-hour monitoring skin electrode, *Proc. 19th ACEMB, San Francisco*, p. 34.

Brown, V. R. and McCusker, D. R., 1968, A new thin film metallized glass microelectrode, *Proc. 21st ACEMB, Houston*, p. 13A5.

Burr, H. S., Hill, R. T., and Allen, E., 1935, Detection of ovulation in the intact rabbit, *Proc. Soc. Exp. Biol. Med.* **33**:109–111.

Caldwell, C. W. and Reswick, J. B., 1967, A new transcutaneous electrode, *Proc. 20th ACEMB, Boston*, p. 15.6.

Delgado, J. M. R., 1964, Electrodes for extracellular recording and stimulation, in *Physical Techniques in Biological Research, Vol. 5* (W. L. Nastuk, ed.), Academic Press, New York.

Ferris, C. D., 1966, Cardiac resuscitation by electronic stimulators, *Final Report USPHS HE–4595*, University of Maryland.

Ferris, C. D. and Cowley, R. A., 1968, Emergency cardiac pacing system, *Proc. 21st ACEMB, Houston*, p. 22A3.

Florey, E. and Kriebel, M. E., 1966, A new suction electrode, *Comp. Biochem. Physiol.* **18**:175–178.

Geddes, L. A. and Baker, L. E., 1968, *Principles of Applied Biomedical Instrumentation*, John Wiley and Sons, New York.

Geddes, L. A., 1972, *Electrodes and the Measurement of Bioelectric Events*, John Wiley (Interscience), New York.

Hagfors, N. R. and Schwartz, S. I., 1966, Implantable electronic carotid sinus nerve stimulators for reducing hypertension, *Proc. 19th ACEMB, San Francisco*, p. 36.

*There are some additional entries which are not cited in the text.

Lagow, C. H., Sladek, K. J., and Richardson, P. C., 1971, Anodic insulated tantalum oxide electrocardiograph electrodes, *Trans. IEEE, BME* **18**(2):162–164.

McMullin, J. J. and Pryor, M. J., 1961, *Proc. 1st Intern. Congr. Metallic Corrosion, London,* p. 51.

Miller, J. and Brooks, C. E., 1971, Chronic percutaneous electronic leads, *Proc. 24th ACEMB, Las Vegas,* p. 202.

Morris, T. W., 1967, Skin-electrode impedance of long term electrodes, *Proc. 20th ACEMB, Boston,* p. 15.5.

Parsons, L., MacMillan, J. H., and Whittaker, J. O., 1958, Abdomino-vaginal electric potential differences with special reference to the ovulatory phase of the menstrual cycle, *Am. J. Obstet. Gynecol.* **75**(1):121–131.

Patten, C. W., Ramme, F. B., and Roman, J., 1966, Dry electrodes for physiological monitoring, *NASA Tech. Note NASA TN D-*3414.

Ray, C. D., Bickford, R. G., Clark, L. C., Johnston, R. E., Richards, T. M., Rogers, D., and Russert, W. S., 1965, A new multicontact, depth probe: details of construction, *Proc. Staff Meet. Mayo Clin.* **40**:771–804.

Richardson, P. C., 1967, The insulated electrode: A pasteless electrocardiographic technique, *Proc. 20th ACEMB, Boston,* p. 15.7.

Richardson, P. C. and Coombs, F. K., 1968, New construction techniques for insulated electro-cardiographic electrodes, *Proc. 21st ACEMB, Houston,* p. 13A1.

Rock, J., Reboul, J., and Wiggers, H. C., 1937, The detection and measurement of the electrical concomitant of human ovulation by use of the vacuum-tube potentiometer, *New Eng. J. Med.* **217**(17):654–658.

Roman, J., 1966, Flight research program—III. High impedance electrode techniques, *Aerospace Med.* **37**:790–795.

Schane, W. P., 1973, Chronic transdermal electrodes, *Proc. 26th ACEMB, Minneapolis,* p. 2.6.

Schaudinischky, L., Moses, N., and Schwartz, A., 1968, New method of measurement of bioelectric potentials, *Proc. 21st ACEMB, Houston,* p. 13A3.

Wise, K. D. and Starr, A., 1969, An integrated circuit approach to extracellular micro-electrodes, *Proc. 8th Int. Conf. Med. Biol. Eng.* Sec. 14-5.

Young, L., 1961, *Anodic Oxide Films,* Academic Press, London.

CHAPTER 9

Signal-Analysis and Filtering Techniques

Signals may be described as functions of time or functions of frequency. In most cases, signals acquired by the use of electrodes are recorded permanently as real-time functions. It is frequently useful to describe signals in terms of their frequency components. Frequency-spectrum information is useful when comparing signals, and when signal-processing equipment is to be used. One can determine amplifier bandwidth necessary to pass the signals with minimum distortion, for example. If a signal is to be filtered to remove noise components, then signal-spectrum information is necessary to prevent filtering of the signal as well as the noise.

This chapter presents certain simple methods for describing mathematically some of the types of signals which are encountered in biomedical research. Techniques are described for both periodic and aperiodic signals. Techniques for extraction of signals from noise are also presented. Signal description and not data reduction is the theme of this chapter. For techniques directed toward statistical analysis of data, the reader is referred to standard reference works (for example, Wortham and Smith, 1959).

9.1. Representation of Complex Periodic Wave Forms

A function $f(t)$ is said to be periodic if

$$f(t) = f(t \pm nT)$$

where T is a constant (the period) and n is any integer. This definition extends for all time, and although mathematically rigorous, it is physically unrealistic. Hence in the context of the experimentalist, a periodic function is one which is repetitive with constant repetition rate and all transient effects associated with the signal have died out.

Certain symmetry relations exist for periodic functions and are tabulated below:

even function	$f(t) = f(-t)$
odd function	$f(t) = -f(-t)$
even-harmonic symmetry	$f(t) = f(t + T/2)$
odd-harmonic symmetry (reflection symmetry about the abscissa, half-wave symmetry)	$f(t) = -f(t + T/2)$
quarter-wave symmetry	$f(t) = -f(t + T/2)$ and $f(t) = \pm f(-t)$

These symmetry relations will be used to advantage in the application of Fourier-series techniques to the description of periodic functions.

9.1.1. The Fourier Series

The Fourier series is a method for describing a complex periodic signal. It affords a representation of the signal as a function of both time and frequency in such a manner that one can easily extract a frequency spectrum for the signal. Since many bioelectric signals may be considered periodic (EKG and EEG wave forms, for example), Fourier-series representation is a natural technique to use, although as will be shown later, other methods may prove more useful when specific data-reduction techniques are to be used.

The Fourier series is one of the general class of trigonometric series described by the expression

$$f(t) = a_0/2 + \sum_{k=1}^{\infty} (a_k \cos k\omega_0 t + b_k \sin k\omega_0 t) \tag{1}$$

where $\omega_0 = 2\pi/T$. The Fourier series is defined when

$$a_k = \frac{2}{T} \int_{-T/2}^{+T/2} f(t) \cos k\omega_0 t\, dt \qquad k = 0, 1, 2, \ldots$$

$$b_k = \frac{2}{T} \int_{-T/2}^{+T/2} f(t) \sin k\omega_0 t\, dt \qquad k = 1, 2, 3, \ldots$$

where $f(t)$ is the description of the signal wave form over one complete cycle (2π radians). Certain restrictions exist concerning the nature of $f(t)$ and the classes of functions to which Fourier-series description is applicable;

Ferris (1961, 1962) gives one method of derivation. Generally speaking, bioelectric signals satisfy the criteria for Fourier-series description. The limiting restrictions are the Dirichlet conditions and are presented here in simplified form:

1. $f(t)$ may have a finite number of discontinuities only over the range of one period.
2. Any infinite discontinuities must be integrable.
3. The integral

$$\int_a^b |f(t)| \, dt < \infty$$

must be absolutely convergent over the range $(a - b)$ of definition of $f(t)$. Normally $(a - b)$ is replaced by $(-T/2$ to $+T/2)$.

A less cumbersome representation can be developed by using Euler's relations:

$$\cos \omega_0 t = \frac{e^{j\omega_0 t} + e^{-j\omega_0 t}}{2}$$

$$\sin \omega_0 t = \frac{e^{j\omega_0 t} - e^{-j\omega_0 t}}{2j}$$

and substituting into Equation (1). This operation yields the complex or exponential representation for the Fourier series.

$$f(t) = \tfrac{1}{2} \sum_{k=-\infty}^{+\infty} c_k \, e^{jk\omega_0 t}$$

where

$$c_k = \frac{2}{T} \int_{-T/2}^{+T/2} f(t) \, e^{-jk\omega_0 t} \, dt \qquad k = 0, \pm 1, \pm 2, \ldots$$

The set of coefficients $\{c_k\}$ is the Fourier spectrum of the function $f(t)$. Alternative formulations are presented in Table 9.1.

The symmetry conditions previously stated affect very strongly the Fourier coefficients a_k, b_k, c_k, d_k. If $f(t)$ is an even function, then all of the coefficients b_k are zero. If $f(t)$ is an odd function, all $a_k = 0$, $k > 1$. Thus an even function has only cosine terms in the series and an odd function has only sine terms. A function with even-harmonic symmetry has only even harmonics in its series (k even). A function with odd-harmonic symmetry has only odd-harmonic terms in its series (k odd). In the two latter instances, both sine and cosine terms may be present. For a function which possesses

TABLE 9.1. Fourier-Series Representations for a Function $f(t)$ or $f(\omega_0 t)$, Period $= T = 2\pi$

Trigonometric forms:

$$f(t) = \frac{a_0}{2} + \sum_{k=1}^{\infty} \left[a_k \cos\left(\frac{2\pi kt}{T}\right) + b_k \sin\left(\frac{2\pi kt}{T}\right) \right]$$

$$a_k = \frac{2}{T} \int_{-T/2}^{T/2} f(t) \cos\left(\frac{2\pi kt}{T}\right) dt \qquad k = 0, 1, 2, \ldots$$

$$b_k = \frac{2}{T} \int_{-T/2}^{T/2} f(t) \sin\left(\frac{2\pi kt}{T}\right) dt \qquad k = 1, 2, 3, \ldots$$

or

$$f(t) = \frac{a_0}{2} + \sum_{k=1}^{\infty} (a_k \cos k\omega_0 t + b_k \sin k\omega_0 t)$$

$$a_k = \frac{1}{\pi} \int_{-\pi}^{\pi} f(\omega_0 t) \cos k\omega_0 t \, d\omega_0 t \qquad k = 0, 1, 2, \ldots$$

$$b_k = \frac{1}{\pi} \int_{-\pi}^{\pi} f(\omega_0 t) \sin k\omega_0 t \, d\omega_0 t \qquad k = 1, 2, 3, \ldots$$

where the substitution $\omega_0 = 2\pi/T$ has been made explicitly

Sine form:

$$f(t) = \frac{a_0}{2} + \sum_{k=1}^{\infty} c_k \sin\left(\frac{2\pi kt}{T} + \theta_k\right)$$

$$= \frac{a_0}{2} + \sum_{k=1}^{\infty} c_k \sin(k\omega_0 t + \theta_k)$$

$$\left.\begin{array}{l} a_k = c_k \sin \theta_k \\[4pt] b_k = c_k \cos \theta_k \\[4pt] c_k = \sqrt{a_k^2 + b_k^2} \\[4pt] \theta_k = \tan^{-1}(a_k/b_k) \end{array}\right\} \quad \text{with } a_k, b_k \text{ defined as shown above}$$

Cosine form:

$$f(t) = \frac{a_0}{2} + \sum_{k=1}^{\infty} c_k \cos\left(\frac{2\pi kt}{T} + \Psi_k\right)$$

$$= \frac{a_0}{2} + \sum_{k=1}^{\infty} c_k \cos(k\omega_0 t + \Psi_k)$$

$$a_k = c_k \cos \Psi_k$$

$$b_k = -c_k \sin \Psi_k$$

$$c_k = \sqrt{a_k^2 + b_k^2}$$

$$\Psi_k = \tan^{-1}(-b_k/a_k)$$

TABLE 9.1. *Continued*

Exponential forms:

$$f(t) = \frac{1}{2} \sum_{k=-\infty}^{\infty} d_k \, e^{j2\pi kt/T}$$

$$d_k = \frac{2}{T} \int_{-T/2}^{T/2} f(t) \, e^{-j2\pi kt/T} \, dt \qquad k = 0, \pm 1, \pm 2, \ldots$$

$$f(t) = \frac{1}{2} \sum_{k=-\infty}^{\infty} d_k \, e^{jk\omega_0 t}$$

$$d_k = \frac{1}{\pi} \int_{-\pi}^{\pi} f(\omega_0 t) \, e^{-jk\omega_0 t} \, d\omega_0 t \qquad k = 0, \pm 1, \pm 2, \ldots$$

---

quarter-wave symmetry, only odd-harmonic cosine terms are present in the associated series. Thus for half-wave symmetry (odd-harmonic)

$$a_k = \frac{4}{T} \int_0^{T/2} f(t) \cos k\omega_0 t \, dt \qquad k \text{ odd}$$

$$= 0 \qquad k \text{ even}$$

$$b_k = \frac{4}{T} \int_0^{T/2} f(t) \sin k\omega_0 t \, dt \qquad k \text{ odd}$$

$$= 0 \qquad k \text{ even}$$

and for quarter-wave symmetry

$$a_k = \frac{4}{T} \int_{-T/2}^{T/2} f(t) \cos k\omega_0 t \, dt \qquad k \text{ odd}$$

$$= 0 \qquad k \text{ even}$$

$$b_k = 0 \qquad \text{all } k$$

As an introductory example, we take the square wave shown in Figure 8.1. It is defined as

$$f(t) = -1 \qquad -T/2 < t < 0 \qquad -\pi/\omega_0 < t < 0$$

$$f(t) = +1 \qquad 0 < t < +T/2 \qquad 0 < t < +\pi/\omega_0$$

By inspection, $a_0 = 0$ since the signal is symmetric with equal area above and below the t-axis for the two half-periods. In addition, the function is odd and possesses half-wave symmetry, thus

$$a_k = 0 \qquad \text{all } k$$

$$b_k = \frac{4}{T} \int_0^{T/2} (1) \sin k\omega_0 t \, dt = \frac{8}{kT}$$

$$= \frac{4}{k\pi} \qquad k \text{ odd (cycle period} = 2\pi \text{ radians)}$$

$$= 0 \qquad k \text{ even}$$

and the Fourier series is simply

$$f(t) = \frac{4}{\pi} \sum_{k \text{ odd}} \left(\frac{1}{k}\right) \sin k\omega_0 t = \frac{8}{T} \sum_{k \text{ odd}} \left(\frac{1}{k}\right) \sin k\omega_0 t$$

$$= \frac{4}{\pi}(\sin \omega_0 t + \tfrac{1}{3} \sin 3\omega_0 t + \tfrac{1}{5} \sin 5\omega_0 t + \dots)$$

$$= \frac{4}{\pi} \sum_{k=1}^{\infty} \frac{\sin (2k - 1)\omega_0 t}{(2k - 1)}$$

The frequency spectrum is plotted in Figure 8.1. For this signal, the phase spectrum is a constant phase angle of 90° if we recall the following:

$$\text{Amplitude frequency spectrum} = \sqrt{a_k^2 + b_k^2} = A(\omega)$$

$$\text{Phase angle spectrum} = \tan^{-1}(b_k/a_k) = \phi(\omega)$$

This result is to be expected since only sine terms appear in the series and there is no sign change.

If we shift the signal shown in Figure 9.1 by a quarter cycle as shown in Figure 9.2, we find a different spectrum function. Again $a_0 = 0$, but the new signal is an even function with quarter-wave symmetry. Thus

$$a_k = \frac{4}{k\pi}(-1)^{(k+3)/2} \qquad \text{for } k \text{ odd}$$

$$= 0 \qquad \text{for } k \text{ even}$$

$$b_k = 0 \qquad \text{for all } k$$

Only cosine terms are present in the series, and the Fourier series is simply

$$f(t) = \sum_{k \text{ odd}} \frac{4}{k\pi}(-1)^{(k+3)/2} \cos k\omega_0 t$$

$$= \frac{4}{\pi} \sum_{k=1}^{\infty} (-1)^{(k-1)} \left| \frac{\cos(2k - 1)\omega_0 t}{(2k - 1)} \right|$$

The amplitude spectrum is the same as before but the phase angle alternates between 0 and 180°.

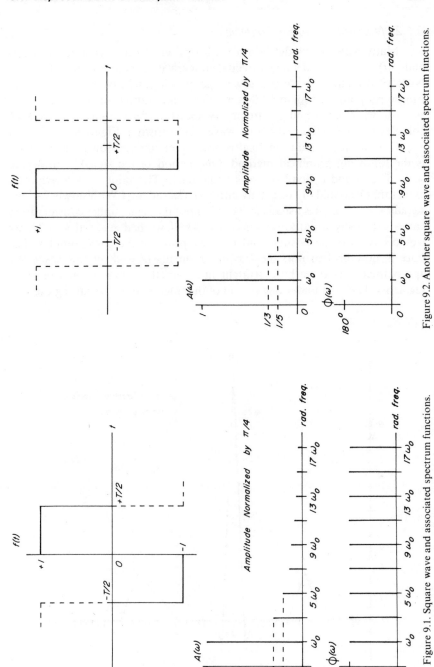

Figure 9.1. Square wave and associated spectrum functions.

Figure 9.2. Another square wave and associated spectrum functions.

9.1.2. Numerical Fourier Analysis

Many wave forms defy simple description. When this situation exists and $f(t)$ cannot be defined directly, numerical techniques must be used. Numerical methods have the advantage of being iterative processes which can be programmed for digital-computer computation. As a case in point, suppose we wish to find the frequency spectrum for the idealized *PQRST* complex of the electrocardiogram wave form shown in Figure 9.3.

One of the simple numerical processes for finding Fourier coefficients is the following graphical method. One period of the signal is drawn to scale. The period is divided into k equal parts. The value for k is arbitrary except that it should divide evenly into 360° (2π radians). With slowly varying signals, $k = 12$ or 24 is usually sufficient. Since the signal under consideration varies extremely rapidly, we select $k = 48$. If we had selected $k = 24$, we would have lost the positive and negative peaks of the *QRS* complex. The trace is now broken into the k equal segments ($k = 48$ in this case). We approximate the trace by a straight line over the range of each of the k segments. The area under one period of the trace is found by taking the sum

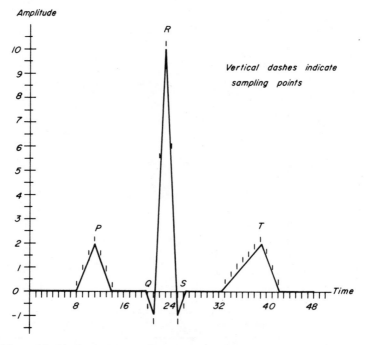

Figure 9.3. Idealized electrocardiogram wave form showing sampling points.

of the areas of the k trapezoids formed by the approximation. For an x–y coordinate system,

$$\text{Area of trapezoid} = \tfrac{1}{2}(y_n + y_{n+1})\,\Delta x = A_n$$

$$\text{Total area under curve (trace)} = \sum_{n=1}^{k} A_n$$

The approximation formulas for the Fourier coefficients are (Sohon, 1944):

$$a_n = \frac{1}{k}[f(0) + f(2\pi)] + \frac{2}{k}\sum_{u=1}^{k-1} f(u\,\Delta x)\cos(nu\,\Delta x)$$

$$b_n = \frac{2}{k}\sum_{u=1}^{u=k} f(u\,\Delta x)\sin(nu\,\Delta x)$$

where

k = number of divisions

u = index of summation (numerical integration)

n = index of desired harmonic

The given signal (EKG wave form) possesses no useful symmetry. Note, however, that $f(0) = f(2\pi) = 0$. Evaluation of the coefficients is most easily achieved by tabulation. Table 9.2 indicates the method. Note that for $k = 48$, a sample point occurs every $7.5°$ in the cycle.

From the table and the formulas, we compute the first few terms of the series:

$$a_0 = \frac{2}{k}\sum_{u=1}^{k-1} f(u\,\Delta x) = (1/24)(32.05) = 1.335$$

$$a_1 = (1/24)(-15.657) = -0.652$$

$$b_1 = (1/24)(-1.109) = -0.046$$

$$f(t) = 1.335 - 0.652\cos\omega_0 t - 0.046\sin\omega_0 t + \cdots$$

where ω_0 is the radian heart rate. If the heart rate is 60 beats/min = 1 beat/sec, then $\omega_0 = 2\pi$ rad/sec. The higher-order coefficients can be evaluated by writing additional tables. It is possible to evaluate coefficients up to a_{24} and b_{23} with the sampling interval chosen.

If u is allowed to range from 0 to k, and if $f(0) = f(2\pi) \neq 0$, then this value should be entered only once in the table to determine the sums $\sum f(u\,\Delta x)\cos(nu\,\Delta x)$. If $f(0) \neq f(2\pi)$, then one-half of each value should be entered in the table.

TABLE 9.2. Fourier Coefficients

u	$u\Delta x$	$f(u\Delta x)$	$\sin(u\Delta x)$	$\cos(u\Delta x)$	$f(u\Delta x) \cdot \sin(u\Delta x)$	$f(u\Delta x) \cdot \cos(u\Delta x)$
0–8	0–60°	0	—	—	—	—
9	67.5°	0.7	0.924	0.383	0.647	0.268
10	75°	1.4	0.966	0.259	1.352	0.363
11	82.5°	2.0	0.991	0.131	1.982	0.262
12	90°	1.4	1.0	0	1.4	0
13	97.5°	0.7	0.991	−0.131	0.694	−0.092
14–20	105°–150°	0	—	—	0	0
21	157.5°	−1.0	0.383	−0.924	−0.383	0.924
22	165°	4.0	0.259	−0.966	1.036	−3.864
23	172.5°	10	0.131	−0.991	1.310	−9.91
24	180°	4.0	0	−1.0	0	−4.0
25	187.5°	−1.0	−0.131	−0.991	0.131	0.991
26–32	195°–240°	0	—	—	0	0
33	247.5°	0.3	−0.924	−0.383	−0.277	−1.149
34	255°	0.6	−0.966	−0.259	−0.580	−0.155
35	262.5°	0.85	−0.991	−0.131	−0.842	−1.114
36	270°	1.1	−1.0	0	−1.1	0
37	277.5°	1.4	−0.991	0.131	−1.387	0.183
38	285°	1.7	−0.966	0.259	−1.642	0.440
39	292.5°	2.0	−0.924	0.383	−1.848	0.766
40	300°	1.3	−0.866	0.500	−1.126	0.65
41	307.5°	0.6	−0.793	0.609	−0.476	0.365
42–48	315°–360°	0	—	—	0	0
Sums		32.05			−1.109	−15.657

It should be noted that this discussion of the Fourier series is incomplete in that matters of convergence and the Gibbs phenomenon are not included. It has been tacitly assumed that any periodic physiological signals encountered in experimental work are accurately represented by some Fourier series. For this reason the detailed mathematical arguments have been omitted for the sake of clarity and brevity (see Carslaw, 1930, for example).

9.2. Frequency Spectra of Aperiodic Functions

Not all bioelectric potentials are periodic. In many cases, single pulse signals are observed, such as neural response to a single stimulus. The Fourier-series method of spectral analysis does not apply in such cases as it is defined for periodic functions only. It can be extended, however, through a series of limiting processes to develop the Fourier transform integrals which are defined below.

$$F(\omega) = \int_{-\infty}^{\infty} f(t) e^{-j\omega t} dt \qquad \text{direct transform spectrum function}$$

$$F(\omega) = \mathscr{F}\{f(t)\} \qquad j = \sqrt{-1}$$

$$f(t) (=) \frac{1}{2\pi} \int_{-\infty}^{\infty} F(\omega) e^{j\omega t} d\omega \qquad \text{inverse transform time function}$$

$$f(t) (=) \mathscr{F}^{-1}\{F(\omega)\}$$

The Fourier transform integrals may be derived directly as well, without using the Fourier series. The first integral yields the frequency spectrum for a given aperiodic function of time $f(t)$. The second integral allows us to find the aperiodic function of time $f(t)$ which corresponds to a given spectral function $F(\omega)$. The conditional equality ($=$) is shown as it is possible to define spectral functions $F(\omega)$ for which no corresponding real time function $f(t)$ exists. To insure direct transformability, $f(t)$ must be a real, aperiodic function of time. The factor $(1/2\pi)$ may be placed with either integral and results from the fact that the integral transform pair has been defined in terms of radian frequency ω instead of natural frequency ($\omega = 2\pi f$). For some purposes, an alternative complex integral definition is used and is shown below, although it will not be used here.

$$F(j\omega) = \int_{-\infty}^{\infty} f(t) e^{-j\omega t} dt$$

$$f(t) (=) \frac{1}{2\pi j} \int_{-j\infty}^{+j\infty} F(j\omega) e^{j\omega t} d(j\omega)$$

This form emphasizes the fact that it is frequently necessary to use contour integration in the complex plane to evaluate the inverse transform integral. In many cases, tables can be used to find both the direct and inverse transforms for given functions $f(t)$ and $F(\omega)$ (Campbell and Foster, 1948). There are a number of basic operations with Fourier transforms which are shown in Table 9.3. Proofs for these operations will be found in the literature (Papoulis, 1962; Ferris, 1962).

Let us now apply the direct Fourier transform to find the frequency spectrum associated with the rectangular pulse shown in Figure 9.4. For convenience, a unit-area pulse has been selected. It is customary to display the aperiodic function in a symmetric manner (as shown) when possible, as this simplifies the expression for $F(\omega)$. This, of course, is subject to the limitations of the physical system involved. The pulse is defined in real time by

$$f(t) = 1/T \qquad -T/2 < t < T/2$$

$$= 0 \qquad \text{all other } t$$

TABLE 9.3. Fourier Transform Operations

Time domain	Frequency domain	Time function	Frequency function
Linearity	Linearity	$\sum\limits_{i} k_i f_i(t)$	$\sum\limits_{i} k_i F_i(\omega)$
Scale change	Scale change	$f(\mu t)$	$\|\mu\|^{-1} F(\omega/\mu)$
Delay	Linear added phase	$f(t \pm \alpha)$	$e^{\pm j\omega\alpha} F(\omega)$
Complex modulation	Spectrum shift	$e^{\pm j\omega_o t} f(t)$	$F(\omega \mp \omega_0)$
Convolution	Filtering	$f_1(t) * f_2(t)$	$F_1(\omega) F_2(\omega)$
Multiplication	Scanning	$f_1(t) f_2(t)$	$F_1(\omega) * F_2(\omega)$
Differentiation	Multiplication by $j\omega$	$d^n f(t)/dt^n$	$(j\omega)^n F(\omega)$
Integration	Division by $j\omega$	$f^{-n}(t)$	$(j\omega)^{-n} F(\omega)$

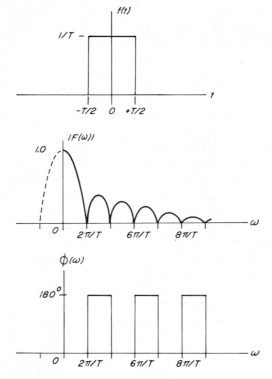

Figure 9.4. Rectangular pulse and associated spectrum functions.

By direct integration we find

$$F(\omega) = \frac{1}{T} \int_{-T/2}^{T/2} e^{-j\omega t} \, dt$$

$$= -\frac{1}{j\omega T}(e^{-j\omega T/2} - e^{+j\omega T/2}) = \frac{2}{\omega T} \frac{e^{j\omega T/2} - e^{-j\omega T/2}}{2j}$$

$$= \frac{\sin(\omega T/2)}{(\omega T/2)} = \frac{\sin x}{x} \qquad x = \omega T/2$$

The amplitude and phase spectra also are shown in Figure 9.4. It should be noted that the defining integrals for $F(\omega)$ and $F(j\omega)$ are identical. In this particular instance, the integration to find $F(\omega)$ yielded a real function of ω. In many instances, the integration yields a complex function in the form shown.

$$F(j\omega) = X(\omega) + jY(\omega)$$

The frequency spectrum function which is plotted is the magnitude of $F(j\omega)$. Thus,

Amplitude spectrum function $= |F(j\omega)| = |F(\omega)| = \sqrt{X(\omega)^2 + Y(\omega)^2}$

and

Phase spectrum function $= \phi(\omega) = \tan^{-1}[Y(\omega)/X(\omega)]$

When $F(j\omega)$ is represented in the form $X(\omega) + jY(\omega)$, it is quite clear how to find $\phi(\omega)$. It is not always clear just how $\phi(\omega)$ will vary if either $X(\omega) = 0$ or $Y(\omega) = 0$. In the case at hand, $Y(\omega) = 0$, but $\phi(\omega)$ is not always equal to zero. This we determine by examining $F(\omega)$. The function $\sin x/x$ changes sign every half-cycle, since this is a property of the sine function. Thus $\phi(\omega)$ alternates between $0°$ and $180°$ as shown in Figure 9.4. This is reasonable when one realizes that $\tan 0° = \tan 180° = 0$.

Figure 9.5 illustrates six additional functions of time and their associated frequency spectra. These functions are characteristic of stimulus-and-response functions associated with biological systems. Further applications of the Fourier transform will be presented in Section 9.3 with regard to filtering of signals to remove noise components.

9.3. Extraction of Signals from Noise

Because of their low signal levels, a major problem which develops in recording bioelectric signals is hum and noise interference. Hum results from 60-Hz pickup from the power mains and from 60- to 120-Hz feedthrough from electronic power supplies. Electrical noise has a variety of

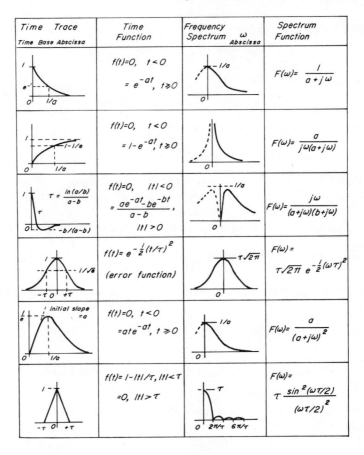

Time Trace Time Base Abscissa	Time Function	Frequency Spectrum ω Abscissa	Spectrum Function
	$f(t)=0, \quad t<0$ $= e^{-at}, \quad t \geqslant 0$		$F(\omega)= \dfrac{1}{a+j\omega}$
	$f(t)=0, \quad t<0$ $= 1-e^{-at}, t \geqslant 0$		$F(\omega)= \dfrac{a}{j\omega(a+j\omega)}$
$\tau = \dfrac{\ln(a/b)}{a-b}$	$f(t)=0, \quad \|t\|<0$ $= \dfrac{ae^{-at}-be^{-bt}}{a-b},$ $\|t\|>0$		$F(\omega)= \dfrac{j\omega}{(a+j\omega)(b+j\omega)}$
	$f(t)= e^{-\frac{1}{2}(t/\tau)^2}$ (error function)		$F(\omega) =$ $\tau\sqrt{2\pi}\, e^{-\frac{1}{2}(\omega\tau)^2}$
Initial slope $=a$	$f(t)=0, \quad t<0$ $= ate^{-at}, t \geqslant 0$		$F(\omega)= \dfrac{a}{(a+j\omega)^2}$
	$f(t)= 1-\|t\|/\tau, \|t\|<\tau$ $=0, \|t\|>\tau$		$F(\omega) =$ $\tau\dfrac{\sin^2(\omega\tau/2)}{(\omega\tau/2)^2}$

Figure 9.5. Typical time-domain functions and their associated frequency functions.

sources. The major ones are radiation from fluorescent lighting installations, ignition and commutator noise. In some environments, 60-Hz signals as high as one volt (peak) can be developed at the input of high-input-impedance oscilloscopes and other instruments.

A first approach to this problem of noise interference is careful shielding of all signal leads, a single common ground point (to prevent ground current loops) with a direct earth ground, and adequate filtering of all dc supply voltages, or use of storage batteries. If vacuum-tube circuits are used, the filaments of all of the tubes should be energized from batteries or a heavily filtered electronic supply. It should be noted, however, that it is possible to develop more heater–cathode hum coupling with a poorly filtered dc supply than with a balanced ac supply.

Many experimental situations, neuron studies for example, require the material under study to be immersed in physiological saline solution. Since saline is a good conductor, it is a potential source of high hum pickup from the electric and magnetic fields produced by the ac power mains. In such cases, either a ground point should be placed in the saline solution (Figure 4.8) or the entire experimental setup should be placed in a shielded screen room.

In most experimental situations, bioelectric potentials are recorded using two electrodes and pickup leads. One lead represents the signal lead and the other the ground return. This is a single-ended or unbalanced pickup system. In this type of system, hum and noise pickup is frequently almost identical in both magnitude and phase in both leads. This is called a common-mode signal since it is common to both leads (see Section 7.5). In the electronics, however, the hum and noise are passed to ground from one lead, while in the other they accompany the desired signal. By using a differential (difference) input amplifier, it is possible to cancel much of the interfering signal. At the amplifier input terminals, the desired signal is unbalanced and is subsequently amplified, while the hum and noise are essentially balanced and cancel. It is necessary to select a differential amplifier which possesses high common-mode-signal rejection. Figure 9.6 presents several circuits for differential amplifiers and illustrates the amount of hum and noise rejection possible with their use.

9.3.1. Filters

If shielding and balanced-signal input techniques are still insufficient to reduce or eliminate undesired signals, then filtering must be considered. One must recognize that in most cases, filtering will cause signal distortion. This is easily demonstrated. Let the composite signal be represented by

$$\phi(t) = s(t) + n(t)$$

where

$$n(t) = \text{noise component}$$

$$s(t) = \text{desired signal}$$

To design a filter to reject $n(t)$, we need to know the spectra of the two signals. This information is obtained by taking the Fourier transform (aperiodic signals) or the Fourier series (periodic signals) of the signal function. For convenience we select the aperiodic case for illustration. Hence

$$\Phi(j\omega) = S(j\omega) + N(j\omega)$$

A filter may be characterized by some transfer function (frequency and phase-angle response) $F(j\omega)$. If we apply the signal $\phi(t)$ to the input of the filter,

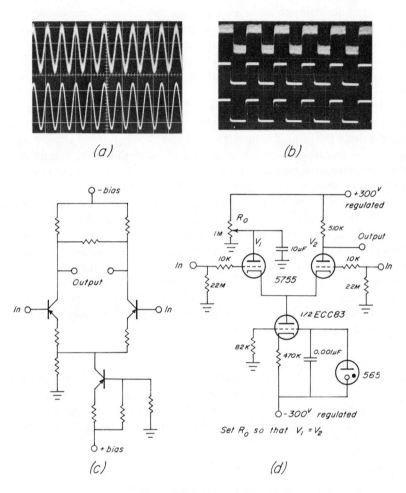

Figure 9.6. Differential amplifier performance with "common-mode" interference signals. (a) Upper trace: 10-kHz sine wave with superimposed common-mode noise as amplified by conventional amplifier; lower trace: same signal amplified by differential amplifier. (b) Upper trace: 10-kHz square wave with superimposed 60-Hz hum and broad-band noise as amplified by single-ended amplifier; middle trace: same signal as amplified by differential amplifier; lower trace: residual noise removed by filtering. (c) Circuit schematic for transistor differential amplifier with constant-current emitter stage. (d) Vacuum-tube differential amplifier in "long-tailed pair" configuration. (See Ferris, 1963.)

the output will be

$$g(t) = \phi(t) * f(t)$$

where $*$ defines a convolution product (see Table 9.3). Substituting in the frequency domain for $\phi(j\omega)$, we find

$$\Phi(j\omega)F(j\omega) = S(j\omega)F(j\omega) + N(j\omega)F(j\omega)$$

Ideally, after filtering, $N(j\omega)F(j\omega) = 0$. This can be achieved only when the amplitude frequency spectrum of the desired signal $|S(j\omega)|$ does not overlap the spectrum of the noise $|N(j\omega)|$. In this case, $F(j\omega) = 1$ over the range of $S(j\omega)$ and $F(j\omega) = 0$ over the range of $N(j\omega)$. In practice, one designs a filter to maximize the signal component $|S(j\omega)F(j\omega)|$ with minimum distortion of $|S(j\omega)|$ and to minimize the noise component $|N(j\omega)F(j\omega)|$.

9.3.2. Simple Filters

Frequently, the noise and signal spectra only partially overlap and relatively simple filters can be used to attenuate the noise. For example, the EKG signal wave form is generally not considered to contain any significant signal components above 40 Hz. EKG recorders are usually equipped with a low-pass filter which cuts off at that frequency. In this manner, 60-Hz hum and high-frequency noise are eliminated from the signal. In cases where the signal has a broad frequency spectrum, with a significant amount of high-frequency information, high-pass filters can be used effectively to eliminate low-frequency noise and hum. When narrow-band signals or narrow-band noise exist, the use of band-pass or band-reject filters respectively is indicated. Figure 9.7 shows configurations for several simple filters.

Hum (60 Hz) in a signal may be greatly attenuated by using a simple RC twin-tee circuit as a rejection filter, provided that the desired signal has no major components at or near 60 Hz. The design conditions for the circuit (Valley and Wallman, 1948) are

$$f_0 = 1/(2\pi RC)$$

$$R' = R/2 \qquad \text{(refer to Figure 9.8)}$$

$$C' = 2C$$

where f_0 is the rejection frequency. Better than 100-to-1 rejection is possible using stable components. Twin-tee filter performance is shown in Figure 9.8.

The slope of the fall-off characteristic of a simple RC filter is 6 db/octave or 20 db/decade. Thus for a 10-to-1 change in frequency there is a 10-to-1 change in signal strength. If simple RC low-pass filters are used to eliminate noise, one must be careful that the cutoff frequency of the filter does not overlap principal frequency components of the desired signal. Because of the

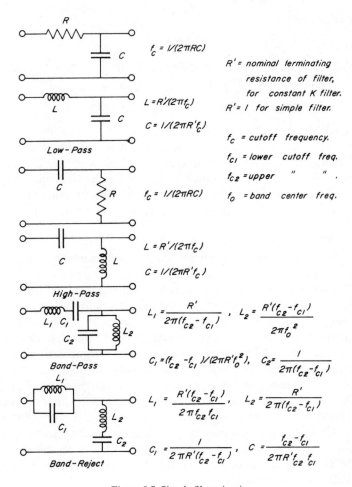

Figure 9.7. Simple filter circuits.

slope characteristic, the *RC* high-pass filter acts as a differentiator in the cutoff region and the *RC* low-pass filter acts as an integrator in its cutoff region.

The transfer function of the *RC* low-pass filter is

$$F(j\omega) = \frac{RC}{j\omega + 1/RC}$$

If $S(j\omega) = 1/(j\omega)$ (a unit step function of voltage) then the filter output signal $g(t)$ will be

$$g(t) = \mathscr{F}^{-1}\{S(j\omega)F(j\omega)\} = 1 - e^{-t/RC} \qquad t > 0$$

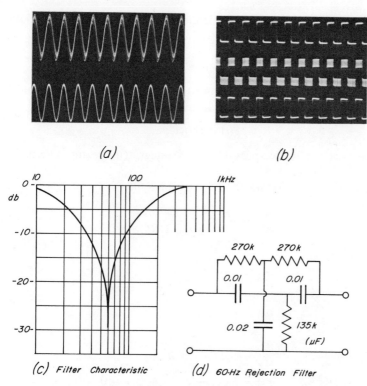

Figure 9.8. 60-Hz twin-tee rejection filter. (a) Upper trace: 10-kHz sine wave with superimposed 60-Hz hum; lower trace: same signal after passing through rejection filter. (b) Upper trace: 10-kHz square wave; middle trace: square wave with superimposed 60-Hz hum; lower trace: same signal after passing through rejection filter. (c) 60-Hz rejection filter frequency response. (d) Filter schematic.

Expanding $g(t)$ in a series we find

$$g(t) = 1 - 1 + \frac{t}{RC} - \frac{1}{2!}\left(\frac{t}{RC}\right)^2 + \frac{1}{3!}\left(\frac{t}{RC}\right)^3 - \cdots$$

$$= \frac{t}{RC} - \frac{1}{2!}\left(\frac{t}{RC}\right)^2 + \cdots$$

The unit step function is defined as

$$u(t) = 1 \qquad t > 0$$
$$= 0 \qquad t < 0$$

The integral of $u(t)$ is the ramp function $u_1(t)$ which has the defining relations

$$u_1(t) = t \qquad t \geq 0$$
$$= 0 \qquad t < 0$$

If RC is large in the preceding expression shown for $g(t)$, we can neglect higher order terms in the series expansion and we see that

$$g(t) = (1/RC)t$$

when $s(t) = u(t)$. The low-pass filter has integrated the input signal and injected a scale factor $(1/RC)$.

The transfer function for the RC high-pass filter is

$$F(j\omega) = \frac{j\omega}{j\omega + 1/RC}$$

Let

$$s(t) = u_1(t)$$

then

$$S(j\omega) = 1/(j\omega)^2$$

$$G(j\omega) = \frac{1}{j\omega(j\omega + 1/RC)}$$

$$g(t) = RC(1 - e^{-t/RC}) \qquad t > 0$$

If RC is small so that $e^{-t/RC}$ approaches $e^{-\infty}$ which approaches zero, then

$$g(t) = RC \qquad t > 0$$
$$= RCu(t)$$

Since the input to the filter was a ramp function $u_1(t)$ and the output from the filter is a step function, the filter has differentiated the signal and applied a scale factor RC. Figure 9.9 illustrates these two conditions.

9.3.3. Fourier Comb Filters

Aside from hum, which is a single-frequency signal, electrical noise is generally a broad-band phenomenon which reflects an impulsive origin. The frequency spectrum of an impulse function (Dirac δ) is given by

$$\int_{-\infty}^{\infty} \delta(t)\, e^{-j\omega t}\, dt = 1$$

All frequencies are present in equal amplitude. We do not find true δ-functions in nature. In the practical case, the amplitude of impulsive noise diminishes

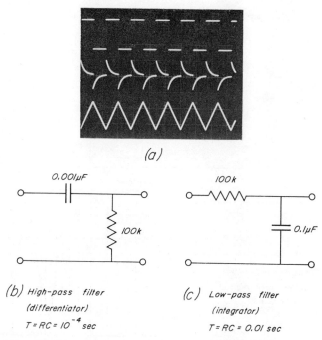

(a)

(b) High-pass filter
(differentiator)

$T = RC = 10^{-4}$ sec

(c) Low-pass filter
(integrator)

$T = RC = 0.01$ sec

Figure 9.9. Distortion introduced by filter networks. (a) Upper trace: 1-kHz square wave; middle trace: differentiated signal output from high-pass filter shown in (b); lower trace: integrated signal output from filter shown in (c), signal period = 1 msec. (b) High-pass filter. (c) Low-pass filter.

as frequency increases. On the other hand, many signals have gaps or null points in their spectra. The square wave, for example, lacks all even harmonic components. A rectangular pulse has a null point every time the spectrum function $(\sin x)/x$ has a null. When the spectral characteristics of a signal are known, one can design filters to reject noise in the regions where the signal spectrum is zero or a minimum.

As an example, let us suppose a signal

$$\phi(t) = s(t) + n(t)$$

is composed of a 5-kHz square wave as shown in Figure 9.10. The noise is broad-spectrum. The spectrum of the square wave is

$$K(\sin \omega_0 t + \tfrac{1}{3} \sin 3\omega_0 t + \tfrac{1}{5} \sin 5\omega_0 t + \cdots)$$

where

$$\omega_0 = \pi \times 10^4 \text{ rad/sec}$$

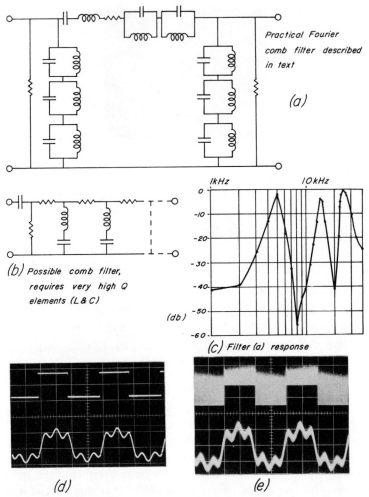

Figure 9.10. Fourier comb filter performance. (a) and (b) Possible circuits. (c) Response of the circuit shown in (a) and discussed in the text; relative gain is expressed in db. (d) Performance characteristics of the filter shown in (a): upper trace is 5-kHz-input square wave; lower trace is filter output for 5-kHz-square-wave input. (e) Filter response under noise conditions; random noise has been added to the 5-kHz-input square wave.

An appropriate filter would consist of a high-pass filter and a series of band-pass filters or band-reject filters. The high-pass filter would be designed to pass from just below 5 kHz and the band-pass filters would be tuned to pass 5 kHz, 15 kHz, 25 kHz, etc. A simple system as shown in Figure 9.10b could be tried. It consists of an *RC* high-pass filter and a series-resonant set of rejection filters in the form of a ladder network. The series-resonant LC circuits are designed to be resonant for the even harmonics of 5 kHz and thus to present a short circuit for all of the even harmonics. This filter then "combs out" noise over selected frequency bands. While such a design works in theory, it will not work in practice because of coupling between LC sections and because unrealistically high Q figures are required for the LC circuits.

A more productive approach is to synthesize a comb to select the frequencies of the signal as proposed initially. Both active and passive filters can be devised. The active filter configuration requires three operational amplifiers in the comb (May and Dandl, 1961) and thus would in general require an inordinate amount of electronics. Recently a method for synthesizing a passive Fourier comb has been reported (Kennedy, 1966). The filter was synthesized for the first three odd harmonics of a square wave using the maximally flat low-pass design proposed by Weinberg (1962). This design was frequency-transformed to be selective at 5, 15, and 25 kHz using Storer's technique (1957). The filter has been duplicated by the author and its performance is shown in Figure 9.10c–e.

9.3.4. Averaging Techniques

If the signal component $s(t)$ of a composite signal

$$\phi(t) = s(t) + n(t)$$

is repetitive with constant repetition frequency, then sampling and averaging techniques may be employed to remove the influence of the noise component $n(t)$. The principle is as follows: Each time the signal trace occurs, it is sampled at a predetermined set of sampling points N. These points occur at the same times relative to each trace for all traces. The sampled values are stored until N' traces have been sampled. The average value of the N' samples for

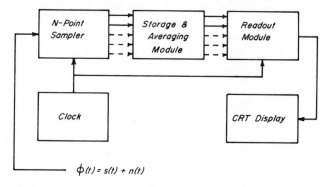

Figure 9.11. System for sampling and averaging to extract repetitive signals from noise.

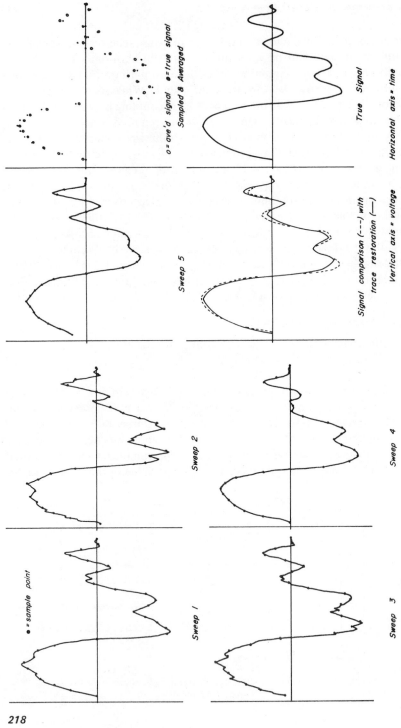

Figure 9.12. Illustration of sampling-and-averaging technique.

each sample point N is computed. The average values are then displayed on an oscilloscope readout. Theoretically, if the signal $s(t)$ and the noise $n(t)$ are uncorrelated, the noise fluctuations should average to zero if enough sweeps N' are used. Typical values are $N = 1024$ sample points per trace sweep and up to $N' = 5000$ sweeps. Figure 9.11 illustrates the basic electronic system for carrying out this operation. The sequences shown in Figure 9.12 show how the process may be carried out manually given a series of N' oscillograph records. In this case, $N' = 5$ and $N = 29$ sample points have been selected. There are 129 voltage levels, from -64 to $+64$ V.

Considering the small number of sweeps ($N' = 5$) and sample points ($N = 29$), the results are quite good, showing that the process of sampling and averaging converges quite rapidly, provided that $s(t)$ and $n(t)$ are uncorrelated.

Commercial units are available to carry out this process, or a general-purpose digital computer can be programmed to perform the operation using magnetic tape input. The limitation of this technique is that $s(t)$ must be repetitive with constant period.

9.3.5. Switching Techniques

Sampling and averaging techniques effect a substantial increase in signal-to-noise level (S/N) for repetitive transients or periodic signals, but are ineffective for single transients. An effective technique exists for processing pulsed signals to reduce the influence of accompanying noise. It uses a gated-switch approach. The output from the processor is a stable baseline unless signal pulses are present. The original system was developed by Plumb and Poppele (1964) and certain refinements were made by Ferris (1966). The system is shown in basic form in Figure 9.13 and in detailed functional form in Figure 9.14. Operation is as follows: The input signal is assumed to be a

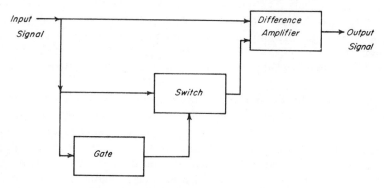

Figure 9.13. Basic signal-from-noise extractor.

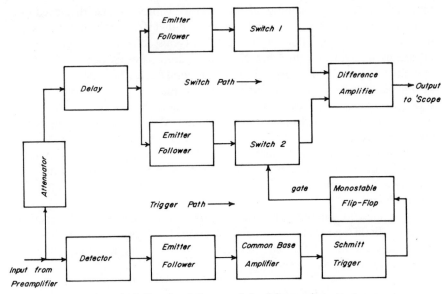

Figure 9.14. Functional diagram of signal-from-noise extractor.

single pulse or a train of pulses of approximately equal duration, but randomly spaced and of random amplitude (above some threshold amplitude). It is also assumed that the signal has passed through a high-gain preamplifier and perhaps has been filtered to some extent.

The signal from the preamplifier is split into three components. The first path leads directly to one input of a difference amplifier. The second path leads through an electronic switch to the other input of the difference amplifier. If the switch is closed, then there is no output from the difference amplifier (dc baseline output), as the difference amplifier takes the difference of two equal signals. The purpose of the gate in the third path is to open-circuit the electronic switch while signal pulses are present. When this happens, the difference amplifier is presented with a signal pulse at one input terminal and no signal at the other input terminal. The output signal is then the amplified signal pulse. The gate is controlled directly by the signal pulses.

In Figure 9.14, the purpose of the various components is as follows: The emitter-follower stages are used for impedance matching; the common-base amplifier is used to amplify the trigger pulse without phase inversion; the difference amplifier has been discussed in the previous paragraph; switch 2 is a silicon transistor switch controlled by the gate; switch 1 is a dummy silicon transistor switch which is always conducting, so placed to insure equal transmission characteristics in both signal paths leading to the difference amplifier; the detector is a simple diode envelope detector used to provide a sharper pulse for the trigger circuitry; the Schmitt trigger and monostable flip-flop provide the gate pulse for the switch (in operation the on time of the monostable flip-flop is adjusted manually to be equal to the duration of the pulses being processed); the delay network is used to hold the signal until the gate has actuated switch 2, and thus compensates for trigger-firing delay.

System performance is indicated in Figure 9.15. The trigger circuit has been adjusted for S/N (voltage) ratio of three-to-one or better. There are two limitations to this system: (1) The signal pulses must be of approximately the same duration (on time for the flip-flop is set for the longest

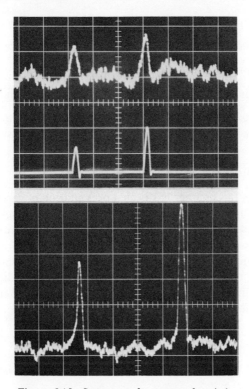

Figure 9.15. System performance of switch-operated signal-from-noise extractor. Upper trace: input signal, 0.5 V/cm. Middle trace: output signal, 2 V/cm. Bottom trace: signals shown above added (using Tektronix type 564 oscilloscope with two-channel preamplifier). The switching circuitry does not distort the pulses. Pulse duration = 1 msec.

duration pulse) and (2) the signal pulses must be of the same polarity. The latter limitation can be removed, however, by designing parallel trigger paths sensitive to signals of opposite polarity. Detailed discussion of this system in regard to circuit design and performance will be found in another publication (Ferris, 1966).

It should be noted that when relatively complex electronic systems are used to process signals, one must take into consideration the noise generated within the system itself. This is generally broad-spectrum noise of thermal origin associated with resistors and has a signal voltage of the form

$$v^2 = 4kT\Delta fR$$

where

$k =$ Boltzmann's constant

$T =$ temperature, °K

$\Delta f =$ system bandpass, Hz

$R =$ effective resistance of system relative to noise generation, Ω

There is also noise produced in vacuum tubes and transistors which varies in voltage as $1/f$ where f is natural frequency. This was discussed in Chapter 7.

9.4. Conclusion

The intent of this chapter has been a presentation of various techniques for representing bioelectric signals, and techniques for processing signals to reduce interference from noise. The techniques presented have been restricted to those which can be most conveniently applied in a laboratory situation. Some of the more sophisticated techniques involving detailed statistical analysis, cross- and auto-correlation techniques, and digital-computer routines have purposely been omitted. Those persons interested in such techniques are directed to the entries listed in the Bibliography.

9.5. References

Campbell, G. A. and Foster, R. M., 1948, *Fourier Integrals for Practical Applications*, D. Van Nostrand, New York.

Carslaw, H. S., 1930, *Introduction to the Theory of Fourier's Series and Integrals*, Dover Publications Reprint, New York.

Ferris, C. D., 1961, Discussion of "Fourier series derivation," *Proc. IRE* **49**:827.

Ferris, C. D., 1962, *Linear Network Theory*, C. E. Merrill, Columbus, Ohio.

Ferris, C. D., 1963, Four-electrode electronic bridge for electrolyte impedance determinations, *Rev. Sci. Instr.* **34**(1):109–111.

Ferris, C. D., 1966, A system for generating a stable dc baseline and suppressing noise when processing signals of physiological and other origin, *Med. Biol. Eng.* **4**(4):381.

Kennedy, E. J., 1966, Passive Fourier comb filter, *Rev. Sci. Instr.* **37**(2):230.

May, F. T. and Dandl, R. A., 1961, *Rev. Sci. Instr.* **32**:387.

Papoulis, A., 1962, *The Fourier Integral and its Applications*, McGraw-Hill, New York.

Plumb, J. L. and Poppele, R. E., 1964, A noise suppressor for neurophysiological recording of impulse activity, *Trans. IEEE PGBME* **11**(4):157.

Sohon, H., 1944, *Engineering Mathematics*, D. Van Nostrand, New York.

Storer, J. E., 1957, *Passive Network Synthesis*, McGraw-Hill, New York.

Valley, G. E., Jr. and Wallman, H., (eds.), 1948, *Vacuum Tube Amplifiers, Vol. 18, MIT Rad. Lab. Series*, McGraw-Hill, New York.

Weinberg, L., 1962, *Network Analysis and Synthesis*, McGraw-Hill, New York.

Wortham, A. W. and Smith, T. E., 1959, *Practical Statistics in Experimental Design*, C. E. Merrill, Columbus, Ohio.

9.6. Bibliography

9.6.1. Signal Representation and Related Subjects

Burch, G. E. and Winsor, T., 1960, *A Primer of Electrocardiography, 4th ed.*, Lea and Febiger, Philadelphia.

Carslaw, H. S., *Introduction to the Theory of Fourier's Series and Integrals, 3rd ed., Revised*, Dover Publications, New York (reprint of 1930 edition).

Carslaw, H. S. and Jaeger, J. C., *Operational Methods in Applied Mathematics*, Dover Publications, New York (reprint of 1948 edition).

Ferris, C. D., 1962, *Linear Network Theory*, C. E. Merrill, Columbus, Ohio.

Scott, R. E., 1960, *Linear Circuits (Part 2)*, Addison-Wesley, Reading, Massachusetts.

9.6.2. Statistical Treatment

Dern, H. and Walsh, J. B., 1963, Analysis of Complex Waveforms, in *Physical Techniques in Biological Research, Vol. 6* (W. L. Nastuk, ed.), Academic Press, New York.

Siebert, W. M. and the Communications Biophysics Group of the Research Laboratory of Electronics, 1959, *Processing Neuroelectric Data*, MIT Press, Cambridge, Massachusetts.

9.6.3. Synthesis of Filter Networks

Ghausi, M. S., 1965, *Principles and Design of Linear Active Circuits*, McGraw-Hill, New York.

International Telephone and Telegraph Corp., *Reference Data for Radio Engineers* (several editions with various dates).

Storer, J. E., 1957, *Passive Network Synthesis*, McGraw-Hill, New York.

Weinberg, L., 1962, *Network Analysis and Synthesis*, McGraw-Hill, New York.

CHAPTER 10

Appendix—Some Practical Matters

Several laboratory procedures and techniques are presented here as an aid and guide for the laboratory or clinical practitioner.

10.1. Measurement of Electrode Tip Size

In order to determine accurately the tip size of most microelectrodes, it is necessary to use an electron microscope because of the diffraction lines produced by light microscopes. With slightly larger electrodes where conventional optical techniques are possible, the effective size (surface area) of a metal electrode may be estimated as follows: A drop of saline solution is placed on a microscope slide and a small wire, connected to the anode of a storage battery, is placed into the drop. The cathode of the battery is connected to the electrode. While this arrangement is being viewed under the microscope, the electrode is slowly placed into the saline. The active area of the electrode is estimated by the surface region from which hydrogen bubbles evolve. Calibrated microscope eyepieces may be used to facilitate area determinations. This method was suggested by Hubel (1950).

10.2. Microelectrode Tip Potential

There appear to be two factors contributing to the tip potential associated with a glass microelectrode. One is independent of electrode tip size and is produced by the liquid junction formed by the electrode filling solution and the electrolyte in which the electrode is placed. The other potential source is related to tip size and the type of glass used. It appears to be related either to ionic gradient at the tip and/or selective permeability of the glass to certain ions in the electrode environment.

The tip potential gives rise to errors in measuring resting and action potentials and can produce overshoot in recording action potentials. It is difficult to measure tip potentials experimentally; however, Schanne *et al.* (1968) have reported that for 0.5-μ-diameter electrodes, it is in the range from 0 to 10 mV. Other studies of tip potential have been conducted by Adrian (1956), Agin and Holtzman (1966), Agin (1969), and Kurella (1969).

10.3. Shielding in EKG Recording

Many clinical electrocardiogram records exhibit a thick tracing as shown in Figure 10.1a. This generally results from improper shielding of the electrode system with concomitant 60-Hz hum pickup. Figure 10.1b illustrates a record obtained from electrodes sutured to the ventricles of a canine heart.

If EKG's are recorded in a Faraday-shielded room using shielded lead wires, records as shown in Figure 10.1c,d can be obtained. Note the absence of 60-Hz interference. In Figure 10.1d, leads I and III exhibit thickening, but this results from muscle tremor signals. The source is the left arm, since this limb is common to both leads. Apparently the strap which anchored the left wrist electrode was too tight, and some EMG signals were produced.

While it is not always possible to record EKG's in a shielded environment, the benefits of doing so, in terms of "clean" traces, are clearly shown in the figure.

10.4. Distortion Produced by Stimulus-Isolation Units

The conventional stimulus-isolation-unit (SIU) method in neurological recording is illustrated in Figure 10.2a and its electric-circuit equivalent in 10.2b. Using Laplace transform notation,

$V_s(s)$ = stimulator voltage output

$V_a(s)$ = voltage applied to physiological system

R_s = series resistance of stimulator output circuit

Z_p = ac polarization impedance at electrode–preparation interface

R, C = circuit model for physiological system

L_1, L_2, M = self- and mutual inductances in the isolation unit

We can solve the circuit shown for the applied voltage $V_a(s)$ as a function of the stimulator output voltage $V_s(s)$ and the system parameters. The result is

$$V_a(s) = Z_T(s)V_s(s)$$

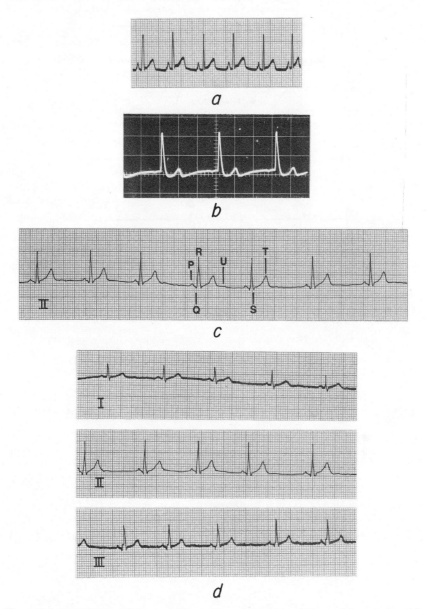

Figure 10.1. Reproduction of electrocardiographic records: (a) Lead II record from canine with 60-Hz trace thickening in record. (b) Record produced by two opposing stainless-steel-mesh disc electrodes sutured to the ventricles of a canine heart (1 mV/cm vertical deflection; time base 0.2 sec/cm). (c) Lead II tracing from young adult human male recorded in a Faraday-shielded room. (d) Leads I–III tracings from young adult human male using shielded leads and recorded in a shielded room. Note EMG signals in leads I and III. Chart speed is 20 mm/sec.

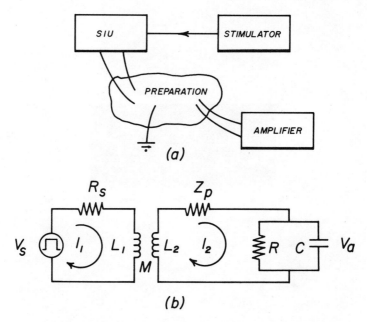

Figure 10.2. (a) Neurological recording system using a stimulus-isolation unit. (b) Electrical circuit model of (a).

where

$Z_T(s) = \pm sMR/D(s)$ (The $\pm$ depends upon the coupling sense of the coils)

$$D(s) = s^3(RCL_1L_2 - RCM^2) + s^2(RR_sCL_2 + L_1L_2 + RCL_1Z_p - M^2)$$
$$+ s(R_sL_2 + RR_sCZ_p + L_1Z_p + L_1R) + RR_s$$

We recognize that $Z_T(s)$ is a nonlinear function as $Z_p = f[s, I_2(s)]$; $s = j\omega$, ω = radian frequency. If we assume that $I_2(s)$ is below the nonlinear threshold for Z_p (Chapter 2) and that perfect coupling exists in the SIU ($M^2 = L_1L_2$), then

$$D(s) = s^2(RR_sCL_2 + RCL_1Z_p)$$
$$+ s(R_sL_2 + RR_sCZ_p + L_1Z_p + L_1R) + RR_s$$

Now even if $Z_p = 0$ (no ac polarization impedance effects implying zero current density but $I_2 \neq 0$), $Z_T(s)$ is still a second-order system function in the form

$$D(s) = s^2(RR_sCL_2) + s(R_sL_2 + L_1R) + RR_s$$

Signal distortion will still occur since

$$V_a(s)/V_s(s) = Z_T(s)$$

For zero signal distortion, the condition that must be satisfied is

$$Z_T(s) = 1$$

In general, $Z_T(s)$ represents a third-order system function. The cubic in s in the denominator $D(s)$ generates at least one real frequency-domain root, which in turn leads to an exponential damping term in the time domain. This causes an increased risetime of the signal applied at the stimulating electrodes.

$$V_a(t) = \mathscr{L}^{-1}\{V_a(s)\}$$

The remaining quadratic factor may have two real (equal or unequal) roots, or a conjugate complex pair of roots, depending upon system parameters. The latter would generate damped time-domain oscillations. These are not usually observed experimentally. The real roots would generate further exponential damping factors that would further degenerate wave-form risetime, or cause "phase tilt" of the pulse top.

Impedance mismatch through the SIU is evaluated roughly by setting $Z_p = 0$ and using the conditions

$$R = R_s, \qquad R > R_s, \qquad R < R_s$$

recognizing that for the isolator, the following conditions may exist:

$$L_1 = L_2, \qquad L_1 > L_2, \qquad L_1 < L_2$$

and that

$$M^2 = L_1 L_2 \qquad \text{or} \qquad M^2 \neq L_1 L_2$$

Thus we have eight interacting boundary conditions. Mathematical analysis is possible, but unwieldy. For a given SIU, the conditions on L_1, L_2 and M are fixed, although generally unknown, unless one determines them experimentally. Figure 10.3 illustrates wave-form distortion introduced by a commercial isolator unit. The system depicted in Figure 10.2 was realized in the laboratory by using a piece of cotton filament saturated with saline solution to simulate an axon. Hence the wave form studied replicates an "artifact" pulse rather than an action potential.

One of the problems associated with using a SIU is loading of the pulse-generator output circuit by a low input impedance of the SIU. This situation is reflected in Figures 10.3a,b. To reduce or prevent this effect, a constant current stimulus source can be used. A conventional pulse generator can be converted to a constant current source by using the circuit of Figure 10.4.

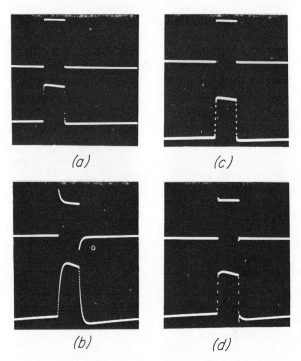

(a)   *(c)*

(b)   *(d)*

Figure 10.3. Oscilloscope records obtained from system shown
in Figure 10.2a. (a) Pulse generator connected to "Hi" terminals
of SIU and preparation connected to "Low" terminals; note
tilt in top of pulse applied to preparation. (b) Pulse generator
connected to "Low" terminals of SIU and preparation connected to "Hi" terminals; note distortion in both wave forms.
(c) Same conditions as (a) but with use of constant current
source. (d) Same conditions as (b) but with use of constant
current source. Note risetime improvement over lower trace
in (b).

The resistors R are made to be the value of resistance into which the generator
is supposed to drive (i.e., 50, 90, 500 Ω, etc.). The operational amplifier forces
the voltage across the load (SIU) to be the same as if the load impedance
were the value that the generator was designed to drive. Figures 10.3c,d
illustrate wave-form improvement when this circuit is used.

10.5. Surface Drying of Preparation

As shown in Figure 10.5a, the integrity of the electrical contact of the
electrodes in a system as shown in Figure 10.2a is of prime importance. The

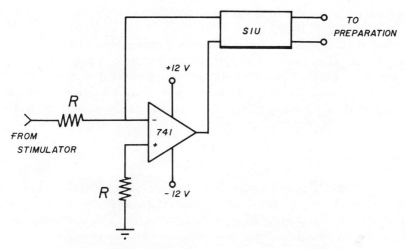

Figure 10.4. Constant current source using a 741 operational amplifier.

grounding electrode is particularly critical. Two factors must be considered. In a standard preparation, the electrodes must be kept wetted with saline to maintain a low-resistance contact. If a mineral-oil preparation is used, electrode contact must be assured before the mineral oil, which is an electrical insulator, is applied. Poor contact, or a contact which is drying out with time,

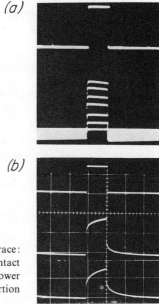

Figure 10.5. (a) Upper trace: stimulating pulse. Lower trace: increase from zero of stimulus artifact as central ground-contact connection dries out. (b) Upper trace: stimulating pulse. Lower traces: amplitude increase (from zero) and wave-form distortion with drying out of electrode contact.

not only causes a change in signal amplitude, but also causes signal distortion as shown in Figure 10.5b.

For these studies, as well as those involving the SIU, a cotton filament soaked with saline was used to simulate the surface of a single nerve fiber or bundle of fibers. The signals studied were simulated "artifact" signals rather than action potentials. There is no loss in generality by using this technique, as the normal "artifact" signal propagates through the electrolyte which wets the preparation surface.

10.6. Suction Electrodes—Precautions

A particular type of suction electrode used in neurological studies was described in Chapter 8. Various gross suction electrodes have been developed for body-surface recording from animals and man. The most commonly used is the precordial suction-cup electrode in electrocardiography.

Several words of caution are warranted in the use of suction-cup surface electrodes. They are easily applied and are popular for this reason in some applications, but prolonged use is contraindicated because they produce a negative pressure gradient at the sites of application. This alters the normal pressure gradient in the capillaries and can cause local injury in the form of bruises. In some uses, injury potentials may occur. These in turn, can mask the desired potential to be recorded.

Since suction electrodes are basically recessed electrodes, problems with electrical contact can develop. One must use a good wetting electrolyte to insure maintenance of a high-integrity electrical connection.

10.7. Surface Impedance Measurement

When two metallic electrodes are placed on the body surface, it is frequently useful to know what electrical impedance (resistance) is presented to these electrodes. There are two analytical approaches to this problem. In either case, we assume uniform electrical conductivity of the biological material. We also assume that the electrodes are small (essentially points with respect to the overall surface).

In a rectilinear volume sample (a cylinder, for example), the resistance between end surfaces is simply

$$R = l/\sigma A$$

where

$$R = \text{measured resistance, } \Omega$$

$$l = \text{sample length, cm}$$

$$A = \text{cross-sectional area of sample, cm}^2$$

$$\sigma = \text{electrical conductivity, mhos/cm}$$

We assume perfectly conducting electrodes and ignore interface effects.

As the first analytical method, we use an electromagnetic theory approach. Assume a body surface of large extent and uniform electrical conductivity σ. Two equal electrodes of radius a are placed on the surface and separated by a distance S, such that $S \gg a$.

Assume a current I enters one electrode and leaves the other electrode. We designate the unit vector $\mathbf{a}_r$ along the surface directed in the positive sense from the electrode at which current enters the surface toward the electrode at which current leaves the surface. We represent the volume which the surface covers as a hemisphere. This is a valid assumption if $a \ll S$, and thus zero potential occurs at a sphere of infinite radius. Now the current density $\mathbf{J}$ (vector sense) developed at the electrode where current enters the surface as measured some radial distance r from the electrode is given by

$$\mathbf{J} = \sigma E = \mathbf{a}_r I/(2\pi r^2) \qquad 0 < r < S$$

where E is the electric field intensity (vector). The contribution of electric field at r from the current-injecting electrode is

$$E_1 = \mathbf{a}_r I/(2\pi\sigma r^2)$$

The contribution of electric field at r from the current-removing electrode is

$$E_2 = \mathbf{a}_r I/[2\pi\sigma(S - r)^2]$$

The potential difference between the electrodes is given by

$$V = \int_a^{S-a} E \cdot \mathbf{a}_r \, dr$$

$$= \frac{I}{2\pi\sigma} \int_a^{S-a} \left(\frac{1}{r^2} + \frac{1}{(S - r)^2} \right) dr$$

Integrating, and evaluating for the indicated limits under the assumption that factors in $1/(S - a)$ vanish (since $S \gg a$) relative to factors in $1/a$, we obtain

$$V = I/(\pi\sigma a)$$

But since the resistance between the two electrodes is

$$R = V/I$$

we obtain as the final result

$$R = 1/(\pi\sigma a) \quad \Omega$$

The final resistance is dependent upon electrode radius a and the medium conductivity σ, but independent of electrode position as long as the electrode separation is large compared with electrode diameter ($S \gg a$).

An alternative approach is to consider the skin surface as an infinite lattice composed of equal resistances R (van der Pol and Bremmer, 1959, pp. 371–372). If we use point electrodes at either end of one resistor, the measured resistance $R' = R/2$. If we measure the resistance across the diagonal of one resistance square of the lattice, $R' = 2R/\pi$. This method does not directly relate electrode size and skin conductivity to resistance values.

10.8. Measurement of Electrode Series Resistance

The technique to be described applies to any electrode that exhibits a high series resistance, but is used mainly for experimental determination of the series resistance of glass micropipettes. Techniques have been described in the literature which use both constant-voltage and constant-current methods under dc or pulse conditions (Frank and Becker, 1964; Geddes, 1972). For dc measurements, provision is made for balancing out the tip potential of a glass micropipette.

A fairly simple general method for determining electrode series resistance is the following: establish a series circuit consisting of an ac voltage source connected to a large indifferent electrode, which in turn is immersed in an electrolyte bath. The electrode whose resistance R_s is to be determined is placed in the bath (the tip of the glass pipette) and then connected in series with a resistor R_L whose value is of the same order of magnitude as the expected value for R_s. The other end of R_L is connected to the voltage source. Let R_e be the electrolyte resistance developed between the indifferent electrode and the tip of the glass pipette. Thus we have a series circuit consisting of V_s, R_e, R_s, and R_L.

Generally R_L and R_s will be of the order of 10^8 Ω. A high-impedance electronic voltmeter with input resistance R_m (from manufacturer's specification sheet) is connected across R_L. Thus the effective resistance at this point in the circuit is R'_L, where

$$R'_L = R_m R_L / (R_m + R_L)$$

Some value of voltage amplitude V_s is selected to establish a current I in the system. This produces a voltage amplitude drop V_L across the resistance R'_L, which registers on the voltmeter. Now

$$I = V_s / (R_e + R_s + R'_L)$$

$$V_L = R'_L I$$

$$= R'_L V_s / (R_e + R_s + R'_L)$$

Normally

$$R_s \sim R'_L$$

$$R_c \lll R_s$$

Thus we can neglect R_c so that

$$V_L = R'_L V_s / (R_s + R'_L)$$

Solving for R_s, we find

$$R_s = R'_L (V_s - V_L)/V_L$$

and finally

$$R_s = \frac{R_m R_L (V_s - V_L)}{(R_m + R_L) V_L}$$

where V_s and V_L are measured voltages, and R_m and R_L are known values of resistance.

It is suggested that a 1-kHz ac voltage be used for this determination to minimize electrode-polarization effects. Stray capacitance should enter into the measurement only as a phase shift in V_L relative to V_s. Since voltage amplitudes are used, this is of no consequence.

10.9. Recessed-Electrode Cell for Impedance Determinations

Figure 10.6 illustrates a cell for use in four-electrode measurements of electrolyte impedance. The working electrodes are platinum. The sensing electrodes are also platinum, but are platinized. They are recessed from the main cell by salt bridges. It is possible to obtain four sample sizes by changing the relative positions of the two sensing electrodes. A detailed description of this cell appears in Schwan and Ferris (1968).

10.10. A Further Note on Signal Distortion by Small Electrodes

In Chapter 4, we discussed signal distortion introduced by the use of microelectrodes. Small-diameter electrodes such as are used in neurological recording (Section 10.4 and Figure 10.2) may introduce similar signal-distortion problems. When the stimulus wave-form risetime is affected, that is, when the risetime is increased, the firing rate of the preparation may be affected, particularly if the preparation is threshold-sensitive. If the preparation manifests a voltage threshold, then artificial delays in firing may be introduced by an increase in the risetime of the stimulus pulse. Errors in conduction-velocity measurements may also result. The use of a constant current stimulator aids in decreasing these problems.

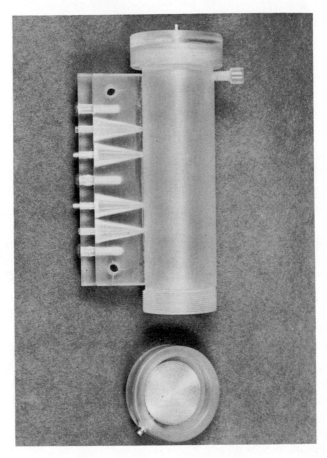

Figure 10.6. Recessed-electrode cell for electrolyte impedance measurements.

In a system as shown in Figure 10.2, the stimulus artifact can be used to estimate Z_p at the recording electrodes. If the signal applied from the stimulus-isolation unit to the stimulus electrodes is visually free from distortion, then any change in wave form of the artifact relative to the stimulus, observed at the recording electrodes, is a function either of contact problems or Z_p at both the stimulus and recording sites. This is under the assumption that the electrolyte sheath around the preparation acts as a simple linear resistor and does not produce signal distortion in itself. Contact can be checked by adding electrolyte at the electrode sites and observing if wave form changes occur. As far as Z_p is concerned, both sites are involved. If similar electrode configurations are used at the stimulating and recording sites, then Z_p may be assumed to be the same at both locations *if current levels are comparable.* The risetime change is then equally distributed between the stimulating and recording electrodes. Thus the *observed* risetime change at the recording electrodes is approximately twice the *actual* risetime change at the recording electrodes. Since the action potential is modified by Z_p at the recording electrodes only, one can estimate the effect of Z_p. If the risetime of the recorded action potential is large relative to the risetime of the recorded artifact, then Z_p is probably of little consequence. If the two risetimes are comparable, then the risetime of the action potential *may* be considerably less than the value recorded. In systems illustrated by Figure 4.10 it is very difficult to estimate Z_p, except by extensive experimentation with the electrode used.

Signal distortion in electrophysiological recording systems may arise from a variety of sources: series ac polarization impedance developed at electrode–electrolyte interfaces; fluid interfaces in glass microelectrode systems and associated fluctuations of tip potentials and related phenomena; stimulus-isolation units; residual and stray capacitances not corrected for by compensating preamplifiers; contact problems; thermal noise and noise generated by other sources.

Pollack (1971) has described signal distortion produced by different types of electrodes used in electromyographic recording. Bipolar electrodes produce differentiation of the recorded signal. Action potentials recorded by a unipolar electrode and an indifferent reference electrode are of the same profile as predicted by theoretical calculations. Coaxial electrodes, as opposed to the bipolar configuration, produce signals which are similar to those recorded from unipolar electrodes operating against an indifferent reference.

10.11. References

Adrian, R. H., 1956, The effect of internal and external potassium concentration on the membrane potentials in frog muscle, *J. Physiol.* **133**:631–658.

Agin, D. P., 1969, Electrochemical properties of glass microelectrodes, in *Glass Microelectrodes* (M. Lavallée, O. F. Schanne, and N. C. Hebert, eds.), John Wiley and Sons, New York.

Agin, D. P. and Holtzman, D., 1966, Glass microelectrodes: origin and elimination of tip potentials, *Nature* **211**:1194–1195.

Ferris, C. D. and Stewart, L. R., 1974, Electrode-produced signal distortion in electrophysiological recording systems, *Trans.IEEE, PGBME* **21**(4):318–326.

Frank, K. and Becker, M. C., 1964, Microelectrodes for recording and stimulation, in *Physical Techniques in Biological Research, Vol. 5* (W. L. Nastuk, ed.), Academic Press, New York.

Geddes, L. A., 1972, *Electrodes and the Measurement of Bioelectric Events*, John Wiley (Interscience), New York.

Hubel, D. H., 1950, Tungsten microelectrode for recording from single units, *Science* **125**:549–550.

Kurella. G. A., 1969. The difference of electric potentials and the partition of ions between the medium and vacuole of the alga, in *Glass Microelectrodes* (M. Lavallée, O. F. Schanne, and N. C. Hebert, eds.), John Wiley and Sons, New York.

Pollack, V., 1971, The waveshape of action potentials recorded with different types of electromyographic needles, *Med. Biol. Eng.* **9**:657–664.

Schanne, O. F., Lavallée, M., Laprade, R., and Gagné, S., 1968, Electrical properties of glass microelectrodes, *Proc. IEEE* **56**:1072–1082.

Schwan, H. P. and Ferris, C. D., 1968, Four-electrode null techniques for impedance measurement with high resolution, *Rev. Sci. Instr.* **39**(4):481–485.

Van der Pol, B. and Bremmer, H., 1959, *Operational Calculus, 2nd ed.*, Cambridge University Press, Cambridge.

General Bibliography

The entries herein present a list of general references and survey papers. Each has been selected for the breadth of coverage and bibliography.

Electrode Processes and Electrochemistry

Adams, R. N., 1969, Electrochemistry at Solid Electrodes, Marcel Dekker, New York.

Bockris, J. O'M. and Conway, B. E. (eds.), 1954–1974, *Modern Aspects of Electrochemistry*, Various imprints: Academic Press, New York; Butterworths, London; Plenum Press, New York.

Bull, H. B., 1971, *An Introduction to Physical Biochemistry, 2nd ed.*, F. A. Davis, Philadelphia.

Conway, B. E., 1965, Theory and Principles of Electrode Processes, Ronald Press, New York.

Damaskin, B. B., 1967, *The Principles of Current Methods for the Study of Electrochemical Reactions*, McGraw-Hill, New York.

Hladik, J. (ed.), 1972, *Physics of Electrolytes, Vol. 1*, Academic Press, London, New York.

MacInnes, D. A., 1961, *The Principles of Electrochemistry*, Reinhold, New York.

Newman, J. S., 1973, *Electrochemical Systems*, Prentice-Hall, New York.

Skoog, D. A. and West, D. M., 1963, *Fundamentals of Analytical Chemistry*, Holt, Rinehart and Winston, New York.

Tobias, C. W. and Delahay, P. (eds.), 1961–1974, *Advances in Electrochemistry and Electrochemical Engineering*, John Wiley (Interscience), New York.

Electronics for Scientists

Brophy, J. J., 1972, *Basic Electronics for Scientists, 2nd ed.*, McGraw-Hill, New York.

Offner, F. F., 1967, *Electronics for Biologists*, McGraw-Hill, New York.

Electrodes: Types, Properties, and Fabrication

Donaldson, P. E. K., 1958, *Electronic Apparatus for Biological Research*, Butterworths, London.

Feder, W. (ed.), 1968, *Bioelectrodes*, New York Acad. Sci. (Annals, Vol. 148, Art. 1), New York.

Geddes, L. A., 1972, *Electrodes and the Measurement of Bioelectric Events*, John Wiley (Interscience), New York.

Hladik, J. (ed.), 1972, *Physics of Electrolytes, Vol. 2*, Academic Press, New York.

Ives, D. J. G. and Janz, G. J., 1961, *Reference Electrodes*, Academic Press, New York.

Lavallée, M., Schanne, O. F., and Hebert, N. C. (eds.), 1969, *Glass Microelectrodes*, John Wiley and Sons, New York.

Nastuk, W. L. (ed.), 1964, *Physical Techniques in Biological Research, Vol. 5*, part A, Academic Press, New York.

Nastuk, W. L. (ed.), 1963, *Physical Techniques in Biological Research, Vol. 6*, part B, Academic Press, New York.

Index